CLINICAL NEUROANATOMY
made ridiculously simple

Stephen Goldberg, M.D.
Professor Emeritus
Department of Cell Biology and Anatomy
University of Miami School of Medicine
Miami, Florida

MedMaster, Inc., Miami

Published by
MedMaster, Inc.
P.O. Box 640028
Miami, FL 33164

ISBN 13: 978-1-935660-19-4
ISBN 10: 1-935660-19-5

To My Parents and to Harriet

CONTENTS

PREFACE

Clinical Neuroanatomy Made Ridiculously Simple is a book intended to help medical students rapidly master that part of neuroanatomy that is essential to clinical care. It is also of practical value to nurses and paramedical personnel who are confronted with neurological problems.

This book was written to fulfill the need for a brief, but readable, summary of clinically relevant neuroanatomy, with examples of medical cases. It is common for neuroanatomy texts to be too greatly oriented toward basic science. They provide far more detail than is necessary to approach clinical situations in neurology. As a result the student becomes confused by mazes of circuitries, often unable to see clinical neuroanatomy as a composite whole and unable to remember essential features.

The student requires two types of books when studying the basic sciences. One is a large, standard reference text which treats the subject as a basic science. *Clinical Neuroanatomy Made Ridiculously Simple* is of the second type, a very small book which focuses directly on the clinically pertinent aspects of that basic science. It is not a synopsis of neuroanatomy; synopses simply condense larger books and are insufficient for the medical student's needs. Smaller, clinically-oriented books must eliminate those aspects of the basic science which have little clinical bearing and emphasize those aspects vital to patient care. Consequently, this book underemphasizes the internal circuitry of the cerebellum, thalamus, and basic ganglia as such knowledge helps little in dealings with neurological problems. However, major organization of the spinal cord and brain stem is strongly emphasized, as this knowledge is vital in neurological localization and diagnosis. The major pathways in the spinal cord are presented simultaneously, rather than in succession, to facilitate comparisons among the pathways. Three of the seven categories of the motor and sensory nuclei have been eliminated in the radically different presentation of the brain stem as a modified spinal cord with only four categories of nuclei: somatic motor, visceral motor, somatic sensory, and visceral sensory.

The mnemonics and humor in this book do not intend any disrespect for patients or original investigators. They are employed as an educational device, as it is well known that the best memory techniques involve the use of ridiculous associations. It is unfortunate that this approach is not attempted more frequently in medical education.

This book is not intended to replace standard reference texts, but rather to be read as a companion text before or during the neuroanatomy course, one which will enable the student to rapidly gain an overall perspective of clinical neuroanatomy. It also provides a rapid review for medical Boards and other exams which tend to emphasize clinically relevant aspects of neuroanatomy.

Rather than text definitions of all potentially unfamiliar terms, a selected glossary follows the text. Clinically oriented questions and answers have also been included, not only for review, but to introduce more subtle information not included in the text.

Dr. Ernst Scharrer, just prior to his untimely death, was my teacher in neuroanatomy at the Albert Einstein College of Medicine. He was a great teacher who could simplify the most complex topics. I am fortunate to have been one of his students.

I am especially indebted to Dr. Ronald G. Clark for his educational influence and encouragement in the preparation of this book. I also thank Drs. Donald Cahill and J. Lawton Smith for their helpful suggestions and Ms. Beryn Frank for editing the manuscript. My neuroanatomy students in the class of 1981 at the University of Miami School of Medicine piloted this book, and I am most grateful for their valuable recommendations. The cover illustration was prepared by Sixten Netzler. Text diagrams are by the author.

PREFACE TO THE INTERACTIVE EDITION

I have been most gratified with the response to *Clinical Neuroanatomy Made Ridiculously Simple* since its initial publication in 1979, and have appreciated the helpful suggestions from readers, and from my students over 25 years of teaching neuroanatomy. A CD-ROM on Neurologic Localization has now been included to enhance the book in several ways:

1. In the "Anatomy" section, clicking on any structure in the nervous system brings up the name of the structure, the neurologic deficit when injured, and graphic explanations for the deficit.

2. In the "Localization Chart," clicking on a patient's symptoms graphically shows the areas of the nervous system that correspond to the deficit, with explanations.

3. The "Lab" section reviews key landmarks in the nervous system, including 3D views.

4. The "Tutorial" section guides the user through the logic in localizing lesions.

5. The "Quiz" section tests the user's ability to accurately localize 40 classic lesions.

I hope the interactive edition will further assist in the application of neuroanatomy to clinical problems.

Stephen Goldberg

CHAPTER 1. GENERAL ORGANIZATION

The central nervous system (CNS) includes the cerebrum, cerebellum, brain stem, and spinal cord (Fig. 1) plus a few scary-sounding structures situated between the brain stem and cerebrum; namely, the *diencephalon* (which includes

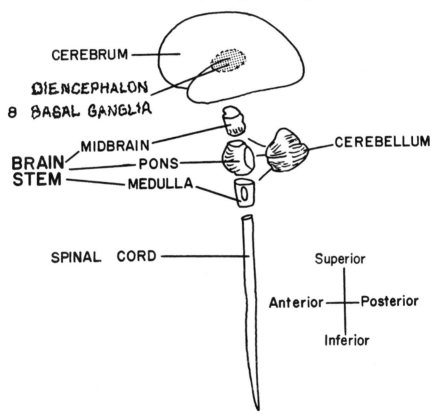

Fig. 1 The central nervous system. Within the brain stem and spinal cord the superior-inferior axis is synonymous with the "rostral-caudal" axis, and the anterior-posterior axis is synonymous with the "ventral-posterior" axis.

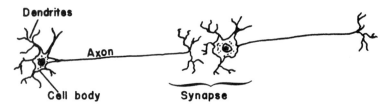

Fig. 2   The neuron.

everything with the name "thalamus;" i.e. the thalamus, hypothalamus, epithalamus and subthalamus) and the *basal ganglia* (which includes the caudate nucleus, the globus pallidus, the putamen, claustrum, and amygdala). Fortunately, it is clinically unimportant to have a detailed understanding of the connections of the diencephalon and basal ganglia. You'll see why later.

The basic functional unit in the CNS is the neuron (Fig. 2). Electrophysiological impulses travel down a neuron from its dendrites to the cell body and axon. Information then is chemically transmitted to other neurons via connections known as synapses. A chain of such communicating neurons is called a *pathway*. Within the CNS, a bundle of pathway axons is called a *tract, fasciculus, peduncle,* or *lemniscus*. Outside the CNS (i.e., in the peripheral nerves, which connect the CNS with the skin, muscles, and other organ systems), bundles of axons are called *nerves*. So you can immediately see the problem with neuroanatomy. There are too many names for the same thing. But the basic logic of neuroanatomy is simple. We shall try to restrict names to a minimum.

There are 31 pairs of spinal nerves and 12 pairs of cranial nerves. Note in figure 3 that cervical nerves C1-C7 exit over their corresponding vertebrae, but that thoracic nerve 1 and the remainder of the nerves exit below their correspondingly numbered vertebrae. Cervical nerve 8 is unique since there is no correspondingly numbered vertebra. Also, note that the spinal cord is shorter than the vertebral column so that the spinal nerve roots extend caudally when leaving the spinal cord. This disparity increases at more caudal levels of the cord. The spinal cord ends at about vertebral level L2 but nerves L2-S5 continue caudally as the *cauda equina* ("horse's tail") to exit by their corresponding vertebrae (Fig. 3).

Figure 4 illustrates the subdivision of the cerebrum into *frontal, parietal, occipital* and *temporal* lobes. These are further subdivided into bulges, called *gyri*, and indentations called *sulci* and *fissures* (small and large, respectively).

The brain stem contains three parts—the *midbrain, pons* and *medulla* (Fig. 1). The pons lies squashed against the *clivus*, a region of bone resembling a slide that extends to the *foramen magnum*, the hole at the base of the skull where the spinal cord becomes the brain stem (Fig. 5).

Sometimes the brain stem does "slide down" the clivus, herniating into the foramen magnum. This is a serious clinical condition, generally resulting from a pressure differential between cranial and spinal cavities. Many clinicians therefore are wary in removing cerebrospinal fluid during a spinal tap in patients with high intracranial pressure.

2

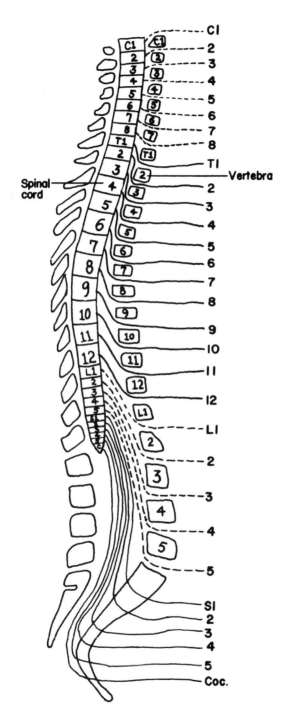

Fig. 3   The spinal nerves. C, cervical; T, thoracic; L, lumbar; S, sacral; Coc., coccygeal nerve.

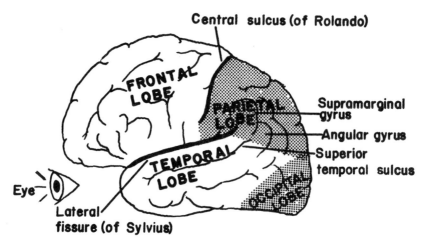

Fig. 4 The cerebrum.

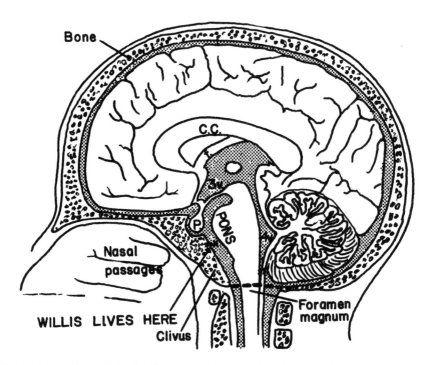

Fig. 5 Sagittal view of the brain. C.C., corpus callosum—the major connection between the two cerebral hemispheres; f, fornix; 3v, third ventricle; p, pituitary gland; 4v, fourth ventricle. Shaded areas are zones containing cerebrospinal fluid.

4

Note the close proximity of the clivus to the nasal passages. Sometimes rare invasive tumors of the nasal passages erode and break through the clivus and damage the brain stem. Pituitary tumors may be reached surgically via the nasal passages by producing a hole in the sphenoidal bone, which houses the pituitary gland—the "transsphenoidal approach".

A spider named *Willis* lives on the pons and its nose fits into the pituitary fossa, but more of this later.

CHAPTER 2.   BLOOD SUPPLY, MENINGES AND SPINAL FLUID

Two main pairs of arteries supply the brain—the two internal carotid arteries and the two vertebral arteries. The vertebral artery changes its name. It's called the basilar artery at the level of the pons and the posterior cerebral artery at the level of the cerebrum (Fig. 6). You'll see why when we discuss Willis, the spider.

Note the important imaginary line in figure 6. It divides the cerebrum into a front (anterior) and a back (posterior) area. The internal carotid artery supplies the front area. Obstruction of the right carotid artery causes weakness and loss of

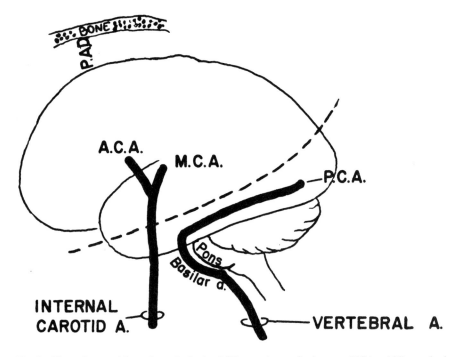

Fig. 6   The major arterial supply to the brain. ACA, anterior cerebral artery; MCA, middle cerebral artery; PCA, posterior cerebral artery; PAD, pia, arachnoid, dura.

sensation on the left side of the body (one side of the brain connects with the opposite side of the body). Blockage of the circulation under the dotted line (vertebral artery distribution) affects the circulation to the visual area of the cerebrum, the brain stem, and the cerebellum and may result in visual loss, dizziness and other problems.

The internal carotid artery divides into an anterior and middle cerebral artery. Note (Fig. 7) that the posterior cerebral artery occupies the entire cerebrum below the dotted line. The middle cerebral artery, though, occupies only the *lateral* surface of the cerebrum above the dotted line, whereas the anterior cerebral artery occupies the entire *midline* area of the cerebral hemisphere above the dotted line.

The brain contains an upside down man named HAL (H-head, A-arm, L-leg), functionally represented on the cerebral cortex. HAL's lower extremity bends over the top of the cerebrum (Fig. 8). Therefore, an occlusion of the anterior cerebral artery results in loss of strength and sensation in the lower part of the body, whereas an occlusion of the middle cerebral artery predominantly affects strength and sensation in the upper regions of the body.

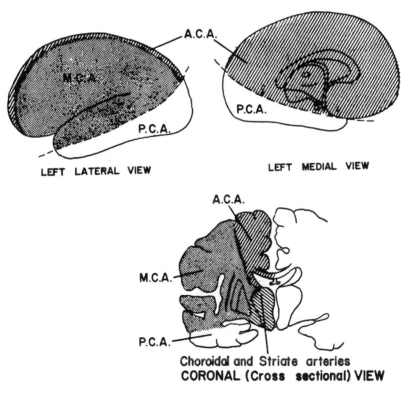

Fig. 7   The cerebral blood circulation. MCA, middle cerebral artery; ACA, anterior cerebral artery; PCA, posterior cerebral artery. (Modified from DeMyer, W., *Technique of the Neurologic Examination,* McGraw-Hill Book Company, 1974.)

One can study the anatomy of the cerebral circulation radiologically by injecting into an artery a contrast material that will outline the blood vessels on x-ray film. This will reveal whether the vessel is blocked or leaking, or of abnormal form or position, resulting from displacement by a tumor or hemorrhage. A catheter (injection tube) threaded retrograde up the right brachial artery to the subclavian artery at the level of the right vertebral artery can be used to release contrast material that will enter both the right vertebral and right carotid arteries, thereby demonstrating the front and back cerebral circulations (Fig. 9). Injection, however, on the left side would demonstrate only the posterior circulation, since the left carotid artery arises directly from the aorta. Thus, the choice of artery and side is important in showing up the desired area in x-ray.

A ferocious spider lives in the brain. His name is Willis! Note (Fig. 10) that he has a nose, angry eyebrows, two suckers, eyes that look outward, a crew cut, antennae, a fuzzy beard, 8 legs, a belly that, according to your point of view, is either thin (basilar artery) or fat (the pons, which lies from one end of the basilar artery to the other), two feelers on his rear legs, and male genitalia. The names in figure 10 look similar to those in figure 6 because they *are* the same structures, seen from different angles. In figure 10 the brain is seen from below, so the carotid arteries are seen in cross section. Figure 10 also explains why the vertebral artery changes its name twice. At first the two vertebral arteries fuse to form one basilar artery. The basilar artery then divides again into two posterior cerebral arteries.

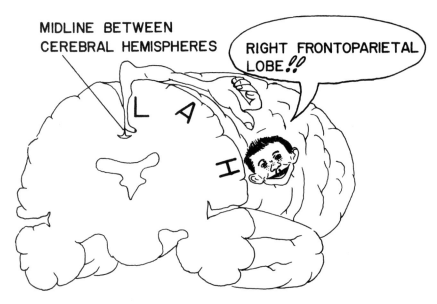

Fig. 8 The homunculus. (Modified from Carpenter, M.B., *Human Neuroanatomy,* The Williams and Wilkins Company, Baltimore, Maryland, 1977.)

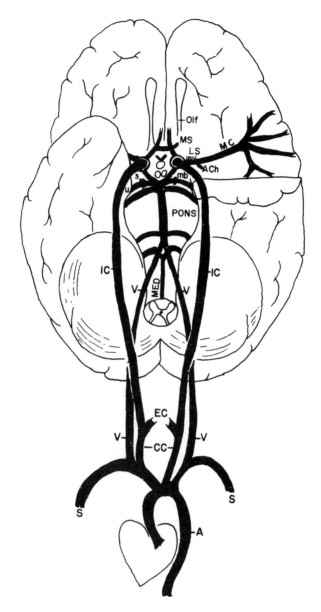

Fig. 9 The origin of the intracranial arteries from the aorta. Olf., olfactory tract; MS, medial striate artery; LS, lateral striate artery; ACh., anterior choroidal artery; MC, middle cerebral artery; 3, cranial nerve 3; mb, midbrain; u, uncus; IC, internal carotid artery; V, vertebral artery; MED, medulla; EC, external carotid artery; CC, common carotid artery; S, subclavian artery; A, aorta. Compare with figures 10 and 25.

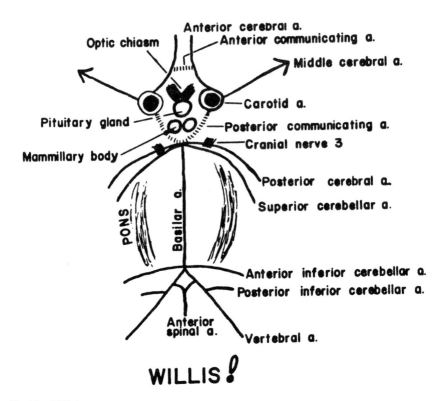

Fig. 10   Willis!

An occlusion of the basilar artery at the junction of the two posterior cerebral arteries will result in total blindness, as the posterior cerebral arteries supply the visual cortex (see Fig. 35). Occlusion of a vertebral artery may result in little or no deficit because of the remaining blood supply from the opposite vertebral artery.

The two communicating arteries are shown as dotted lines in figure 10 because blood flow shows no particular tendency to go one way or the other along these channels. This is logical since blood normally flows up both the carotid and vertebral arteries, equalizing the pressure on both sides. Hence, contrast material injected into the right carotid artery generally will not cross over to the left side of the brain via the anterior communicating artery or flow back into the basilar artery across the posterior communicating artery. This all goes to show that the brain is smart. If one of the major vessels is occluded, the communicating arteries function as anastomoses.

Question: Which subclavian artery would you inject with contrast to demonstrate both the carotid and vertebral circulations?

Ans: Right.

Question: Contrast injection into the left vertebral artery shows up which side of the brain, left or right?

Ans: Both sides; contrast enters the basilar artery and then both posterior cerebral arteries.

Willis has hairy armpits—the third cranial nerve exits between the posterior cerebral artery and the superior cerebellar artery. An *aneurysm* (a weakness and focal ballooning out of the wall of a blood vessel) which affects either of the above two blood vessels may press upon and damage the third nerve.

There are an anterior, middle, and posterior cerebral artery. It would have been nice to have an anterior, middle and posterior cerebellar artery, too, but someone inconsiderately named these three arteries differently.

It could have been worse. He could have named them after himself—actually, he did, for he was a real SAP (S-superior cerebellar artery, A-anterior inferior cerebellar artery, P-posterior inferior cerebellar artery). The cerebellar arteries supply not only the cerebellum but also parts of the brainstem. Their occlusion will result in damage to corresponding areas of the brain stem.

The Veins

Unlike other arteries of the body which have corresponding veins, Willis has no female counterpart. This is because he is so ugly that the veins flee in the opposite direction, jumping clear out of the brain and directly into the *dural sinuses*. You see, the brain is separated from the cranial bone by a PAD (P-pia, A-arachnoid, D-dura membranes—Fig. 6), otherwise known as the meninges. The pia is thin and vascular (Willis lives in it) and hugs the brain. The arachnoid lies between the pia and dura, is arachnoid-like, like a gossamer spider web, and avascular. The dura lies up against the bone and is thick and durable, containing a double layer of connective tissue (the outer layer representing periosteum) with thick venous channels, called sinuses, that lie between the two layers. The dura dips down in between the cerebral hemispheres as the *falx cerebri* and in between the cerebrum and cerebellum as the *tentorium cerebelli*. The "PAD" surrounds the entire central nervous system, including the spinal cord and optic nerve.

The term "supratentorial" has been applied notoriously in a clinical sense, to imply that a patient's problem is hysterical, and not organic. I.e., "His abdominal pain is supratentorial," above the tentorium and "in his head," and not in the abdomen where the pain is reported.

Three main sinuses (Fig. 11) deserve mention: 1) Spinal fluid drains into the *superior sagittal sinus*. 2) The *cavernous sinus*, into which venous blood drains from the eye, provides a potential source of entry into the brain of orbital and facial infections. Many important structures run by or through the cavernous sinus, including the carotid artery and all the nerves entering the orbit. These can be damaged by disease in this critical area. Sometimes the carotid artery breaks open into the cavernous sinus (a "fistula") resulting in blood backing up into the orbital veins and a protruding eye with dilated blood vessels. 3) The *transverse sinus* runs by the ear and may become involved in inner ear infections (Fig. 11).

The veins of the brain drain into the internal jugular vein, a structure you know well from vampire movies.

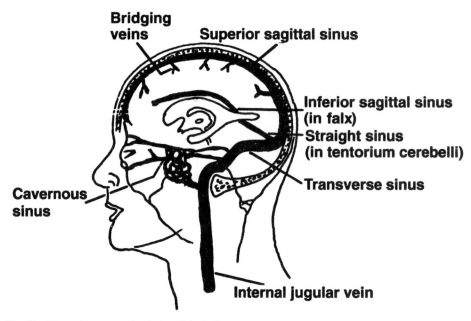

Bridging veins

Superior sagittal sinus

Inferior sagittal sinus (in falx)

Straight sinus (in tentorium cerebelli)

Transverse sinus

Cavernous sinus

Internal jugular vein

Fig. 11   The major venous circulation of the brain.

The Cerebrospinal Fluid

The brain and spinal cord are hollow—which is fortunate for that means there is less to know (Fig. 12). Various points in this fluid-filled hollow tube are expanded and termed *ventricles*. The walls of each ventricle contain a specialized structure called the *choroid plexus* which secretes the clear, colorless *cerebrospinal fluid (CSF)*. CSF flows from the two *lateral ventricles* through the *two interventricular foramina* (holes), first into the single mid-line *third ventricle*, from there through the single midline *aqueduct of Sylvius*, next into the single midline *fourth ventricle*, and then passes outside the brain via three openings in the fourth ventricle: a *M*iddle foramen of *M*agendie and two *L*ateral foramena of *L*uschka. Once outside the brain stem, the CSF enters the *subarachnoid space* (the space between the arachnoid and pia) and exits into the superior sagittal sinus via specialized structures in the sinus wall called *arachnoid villi*. An obstruction at any point along this pathway will lead to dilation (expansion) of the lateral ventricles, termed *hydrocephalus*.

Normally, CSF does not circulate in the central spinal canal of adults as the canal is closed off in places.

Expanded areas of the subarachnoid space are called *cisterns*, the largest of which is the lumbar cistern, between vertebrae L2 (where the spinal cord ends) and S2 (Fig. 12) in adults. Spinal fluid is extracted from this space during the "spinal tap." Do not panic on doing a spinal tap. There is plenty of room, some five vertebral interspaces, and the crest of the iliac bone provides a landmark to

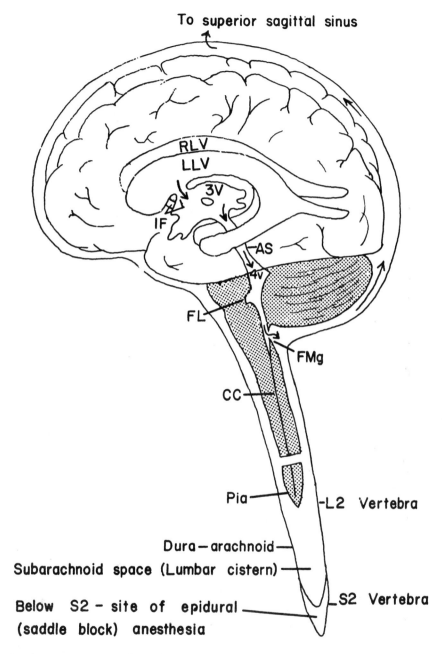

Fig. 12   The cerebrospinal fluid (CSF) circulation. Arrows indicate the direction of flow of CSF. RLV, right lateral ventricle, LLV, left lateral ventricle; 3V, third ventricle; IF, inter-ventricular foramina; AS, aqueduct of Sylvius; 4V, fourth ventricle; FL, foramen of Luschka, FMg, foramen of Magendie, CC, central canal.

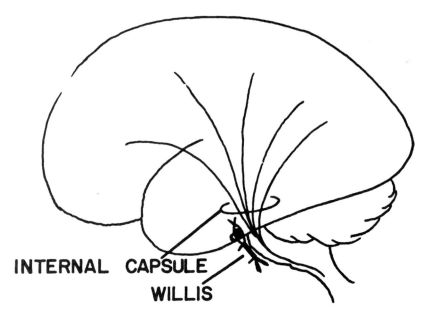

INTERNAL CAPSULE

WILLIS

Fig. 13   Willis and the internal capsule.

determine approximately the center of the lumbar cistern. There is little chance of hitting the spinal cord and it is uncommon to damage the spinal roots of the cauda equina. In newborns, the anatomy differs, the spinal cord ending (*conus medullaris*) at about vertebra L3 rather than the adult level of L1 or L2, ascending as the child grows.

The CSF flows freely throughout the subarachnoid space; thus a *subarachnoid hemorrhage*, usually the result of leakage from an aneurysm in the region of Willis, is felt as a severe neckache or backache, reflecting the seepage of blood down to spinal cord regions. A spinal tap will reveal such bleeding. Symptoms in meningitis are similar, because of the diffusion of pus cells down the subarachnoid space.

A *subdural hemorrhage* results from tearing the bridging veins (Fig. 11) that connect the brain and dural sinuses. In an *epidural hemorrhage*, blood collects between the dura and bone, the result of tearing arteries, particularly the *middle meningeal artery* which lies outside the dura and forms a groove in the cranial bone. Epidural hemorrhage generally coincides with a skull fracture. Large sub-dural and epidural hemorrhages compress the brain and may cause significant damage.

It is possible to suffer a major stroke either by a massive arterial occlusion or by an occlusion of a very small artery situated in a critical area. For instance, a sudden occlusion of the internal carotid artery may cause a massive cerebral infarction. An occlusion of one of the tiny arteries that arise from Willis' head (*medial striate, lateral striate,* or *anterior choroidal arteries*—Figs. 7, 9) may cause as much deficit by infarcting the *internal capsule*. The internal capsule is the narrow zone of a funnel of motor and sensory fibers that converge upon the brain stem from the cerebral cortex (Fig. 13). Each cerebral hemisphere contains one internal capsule, situated deep within the brain just behind Willis' head. The anterior choroidal artery and striate arteries supply the internal capsule and may hemorrhage in situations of hypertension or arteriosclerosis, thereby causing much damage from a tiny lesion.

Questions

2-1 An occlusion of which area of the circle of Willis will result in total unilateral blindness?
Ans. An occlusion of the ophthalmic artery, which is a branch of the internal carotid artery.

2-2 In injecting the right carotid artery in an angiogram how might one simultaneously fill the circulation in the left anterior and middle cerebral arteries?
Ans. Simultaneously compress the left carotid artery during injection. The pressure differential enables blood to pass to the left circulation via the anterior communicating artery.

2-3 Skull fracture is to epidural hemorrhage as aneurysm is to_____?
Ans. Subarachnoid hemorrhage.

2-4 In the region of the lumbar cistern, which membrane is in closest apposition to the arachnoid membrane—the dura or pia?
Ans. The dura. The pia always lies apposed to the brain and spinal cord. The lumbar cistern is a large subarachnoid space and the arachnoid is next to the dura.

CHAPTER 3.   SPINAL CORD

The spinal cord, as seen in cross section (Fig. 14), contains central *grey matter* and peripheral *white matter*. The grey matter contains many neuronal cell bodies and synapses. The white matter contains ascending and descending fiber pathways. The ascending pathways relay sensory information to the brain. The descending pathways relay motor instructions down from the brain. The number of synapses within a pathway is not very important to know clinically. However, it is vital to remember the sites of contralateral crossing over of the various pathways and also the location of the pathways within the white matter as seen in cross section (Fig. 14).

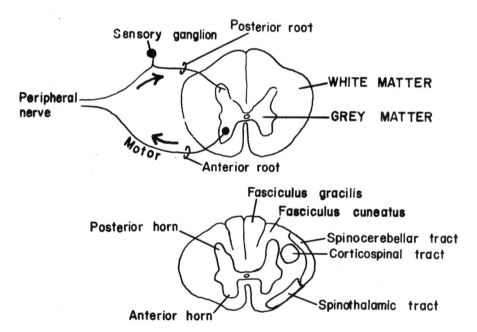

Fig. 14   Major subdivisions of the spinal cord. The posterior columns consist of the fasciculus gracilis plus fasciculus cuneatus.

The Sites of Crossing Over

There are 3 main sensory systems entering the spinal cord:

1. Pain-temperature

2. Proprioception—stereognosis (see glossary)

3. Light touch

To avoid confusion, no sensory synapses will be drawn in figure 15. As a rule though, there is a tendency for sensory fibers to synapse just prior to crossing over (Fig. 16).

The pathway for pain and temperature enters the spinal cord, crosses over to the opposite half of the cord *almost immediately* (actually within one or two spinal cord vertebral segments), ascends to the thalamus on the opposite side, and then moves on to the cerebral cortex. In reality many fibers end in the brain stem and never make it as far as the thalamus, but these will not concern us here. A lesion of the *spinothalamic tract* will result in loss of pain-temperature sensation contra-laterally, below the level of the lesion.

The pathway for proprioception and stereognosis (also for the perception of vibration) initially *remains on the same side* of the spinal cord that it enters, *crossing over* at the junction between the spinal cord and brain stem. The synaptic areas just prior to this crossing are the *nucleus cuneatus* and *nucleus gracilis* (Fig. 16). Nucleus gracilis conveys proprioceptive information from the lower part of the body, whereas nucleus cuneatus conveys information from the upper levels. Their corresponding spinal cord pathways are termed *fasciculus gracilis* (graceful, like

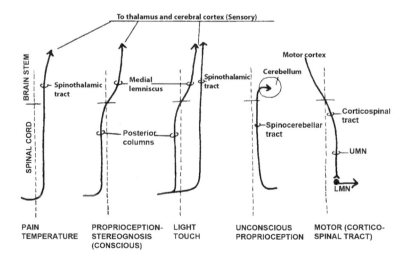

Fig. 15 Schematic view of the major ascending and descending pathways. UMN, upper motor neuron; LMN, lower motor neuron.

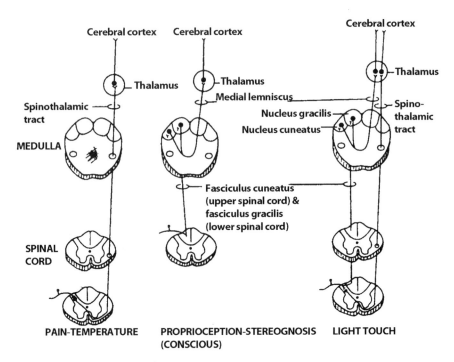

Fig. 16   Three major sensory pathways.

a ballerina's legs) and *fasciculus cuneatus* (cunning, being closer to the brain). The fasciculus gracilis and fasciculus cuneatus are collectively termed the *posterior columns* (Fig. 14). A lesion of the posterior columns results in a decrease in conscious proprioception and stereognosis (also vibration sense) ipsilaterally below the level of the lesion. (Actually, the deficit is mostly in stereognosis, since conscious proprioception and vibration sense are to some extent represented in more lateral regions of the cord.)

The path for light touch combines features of the above two pathways. It partly remains uncrossed until it teaches the level of the brain stem, and partly crosses over at lower levels. This is why light touch typically is spared in unilateral spinal cord lesions; there are alternate routes to carry the information.

All the above sensory pathways eventually cross over and terminate in the thalamus. From there, they are relayed to the sensory area of the cerebrum. A lesion of the sensory area of the cerebral cortex may result in a contralateral deficit of all the above sensory modalities.

Proprioception-stereognosis has a conscious and an unconscious component. The conscious pathway, mentioned above, connects with the thalamus and cerebral cortex, enabling you to describe, for instance, the position of your limb. The unconscious pathway connects with cerebellum (the cerebellum is considered an unconscious organ) as the *spinocerebellar pathway* and enables you to walk and perform other complex acts subconsciously without having to think which

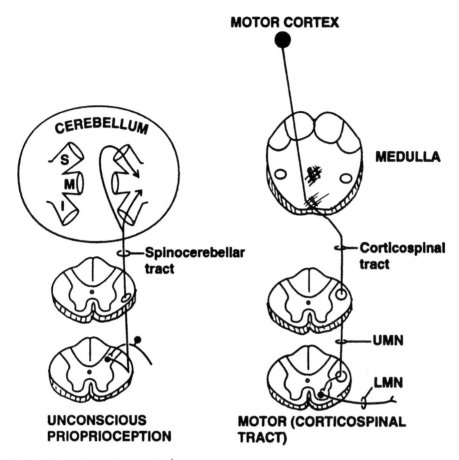

MOTOR CORTEX

CEREBELLUM

S
M
I

MEDULLA

—Spinocerebellar
tract

—Corticospinal
tract

—UMN

—LMN

**UNCONSCIOUS
PRIOPRIOCEPTION**

**MOTOR (CORTICOSPINAL
TRACT)**

Fig. 17 The spinocerebellar and corticospinal tracts. S., superior cerebellar peduncle; M., middle cerebellar peduncle; I., inferior cerebellar peduncle. UMN, upper motor neuron; LMN, lower motor neuron.

joints are flexed and extended (Fig. 17). (In this book where it is stated that a patient presents with "proprioceptive loss," this refers to loss of *conscious* proprioceptive sensation, as tested by asking the patient to describe the position of a limb.)

Unlike the other sensory pathways, which cross contralaterally, the spinocerebellar primarily remains ipsilateral. In general, one side of the cerebellum connects with the same side of the body. Thus *cerebellar lesions tend to produce ipsilateral malfunctioning, whereas cerebral lesions result in contralateral defects.*

There are three main connections between the cerebellum and the brain stem— *the superior, middle, and inferior cerebellar peduncles*, which connect with the midbrain, pons, and medulla, respectively. The spinocerebellar pathways enter the cerebellum via the superior and inferior peduncles (Fig. 17).

The motor (*corticospinal*) pathway is relatively simple. It extends from the motor area of the cerebral cortex down through the brain stem, crossing over at approximately the same level as the *medial lemniscus* (at the junction between brain stem and spinal cord) (Figs. 15, 17). (There is an uncrossed component that will not concern us here.)

The corticospinal pathway synapses in the anterior horn (motor grey matter) of the spinal cord just prior to leaving the cord. This is important, for motor neurons above the level of this synapse (connecting the cerebral cortex and anterior horn) are termed *upper motor neurons* (UMN), whereas those beyond this level (the peripheral nerve neurons) are termed *lower motor neurons* (LMN). These terms are actually somewhat misleading and probably would better have been phrased as first order motor neurons (UMN) and second order motor neurons (LMN), as either of these categories may have axons that lie in the upper or lower part of the body. Upper and lower motor neuron injuries result in different clinical signs. Although both result in paralysis, they differ as follows:

Upper MN Defect	*Lower MN Defect*
spastic paralysis	flaccid paralysis
no significant muscle atrophy	significant atrophy
fasciculations and fibrillations not present	fasciculations and fibrillations present
hyperreflexia	hyporeflexia
Babinski reflex may be present	Babinski reflex not present

Cross Sectional Location of the Pathways

Figure 14 illustrates the positions of the sensory and motor pathways in cross section.

For unknown reasons, various diseases affect different portions of the spinal cord. *Amyotrophic lateral sclerosis* (Fig. 18A) is characterized by a combination of upper and lower motor neuron signs—muscle weakness, atrophy, fibrillations, and fasciculations combined with hyperreflexia. The lesion involves both the anterior horns (motor) of the grey matter (causing a lower motor neuron lesion) as well as the corticospinal tracts (resulting in an upper motor neuron lesion).

Tertiary syphilis (tabes dorsalis) includes proprioceptive loss and pain (posterior root irritation), particularly affecting the lower extremities. The lesion involves the posterior columns of white matter (Fig. 18B) and may extend to the posterior roots and root ganglia.

In *pernicious anemia* (Fig. 18C), there is proprioceptive loss and upper motor neuron weakness. The lesion involves the posterior columns and corticospinal tracts.

Polio attacks the anterior horn cells (Fig. 18D) leading to lower motor neuron involvement, with weakness, atrophy, fasciculations, fibrillations, and hyporeflexia.

Patients with *Guillain-Barre syndrome* (Fig. 18E) experience sensory and lower motor neuron loss because of peripheral nerve involvement.

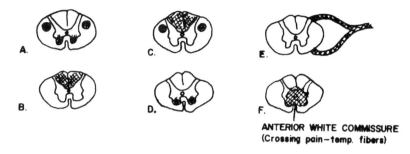

A.

B.

C.

D.

E.

F.

ANTERIOR WHITE COMMISSURE
(Crossing pain−temp. fibers)

Fig. 18   Lesions sites in classical diseases. A. Amyotrophic lateral sclerosis (Lou Gehrig's disease). B. Tertiary syphilis. C. Pernicious anemia. D. Polio. E. Guillain-Barre. F. Syringomyelia.

Certain muscle groups may be affected more than others in an upper motor neuron lesion. In the typical cerebral hemisphere stroke secondary to carotid artery occlusion, the patient develops *decorticate posturing*, characterized by flexion of the wrist and elbow and extension (straightening) of the ankle and knee. In midbrain strokes, the posturing is similar, except that the elbow is extended (*decerebrate posturing*). The mechanisms apparently involve injury to other motor pathways (extrapyramidal) outside the corticospinal (pyramidal) system. "Pure" lesions of the corticospinal tract result predominantly in difficulty with skilled movements of the distal aspect of the extremities.

The Autonomic (Visceral) Nervous System

The pathways mentioned above belong to the *somatic motor* and *somatic sensory* systems. Somatic motor fibers innervate striated skeletal muscle. Somatic sensory fibers innervate predominantly the skin, muscle and tissues other than the viscera. "*Viscera*" means cardiac muscle, smooth muscle (as in the gut) and glands. Visceral (autonomic) motor nuclei are located in between the somatic sensory and motor areas of the spinal cord grey matter. It is useful to think of the autonomic system as consisting of visceral sensory and visceral motor components (Fig. 19).

The chemical and functional differences between sympathetic and parasympathetic autonomic fibers will be discussed in Chapter 6. For the present, simply note that sympathetic motor fibers from the spinal cord synapse relatively near the spinal cord, whereas parasympathetic motor fibers synapse very close to or within the end organs (Fig. 45).

Questions

3-1   A hemisection of the spinal cord at the level of T1 actually produces contralateral loss of pain-temperature sensation of T3 and below, rather than at T1 and below. Why?
Ans.   Actually, pain-temperature fibers do not all cross over immediately. Many ascend 1 or 2 segments before crossing over.

The pain-temperature fibers in the spinothalamic tract are arranged so that Leg fibers are lateral to Arm fibers. Hence, a tumor pressing in from outside the cord may affect the lower extremities fibers first. In such case contralateral pain loss may begin many segments below the lesion, ascending with time as the tumor grows.

3-2   The anterior spinal artery (Fig. 10) extends down the spinal cord, supplying the anterior two-thirds of the cord. Two *posterior spinal arteries* come off variably from the posterior inferior cerebellar arteries or vertebral arteries and supply the rest of the spinal cord along with anastomoses from *intercostal and other arteries*. What sensory modalities might you expect to be preserved following infarction of the anterior spinal artery at spinal cord levels?
Ans.   Proprioception-stereognosis and light touch (posterior columns).

3-3   Locate the lesion. The patient has bilateral paralysis, fasciculations and muscle atrophy at the level of C8T1 along with bilateral pain-temperature loss at the level of T1T2.
Ans.   *Syringomyelia*—a degenerative disease of the central cord or brain stem of unknown cause, affecting the crossing pain-temperature fibers. Depending upon the size and shape of the lesion, which may be asymmetrical, or, as in this case, symmetrical, other areas of the cord may be involved, such as the motor horns, in this case at about level C8T1 (Fig. 18F).

3-4   Surgical tractotomy—severing the spinothalamic tract to relieve pain often affords only temporary relief. Why?
Ans.   Possibly other pathways of unclear nature may take over. Severed axons within the human central nervous system characteristically regenerate poorly.

3-5   At what level of the nervous system—peripheral nerve, spinal cord, or above—would you expect a lesion to lie, with the following signs?
A. Sensory loss along a given *dermatome*, unilaterally.
Ans.   Peripheral nerve or its entry point into the spinal cord. Generally, because of the arrangement of fibers in the ascending pathways and cerebral cortex, it is unlikely that only a single dermatome will be affected by a given lesion in the spinal cord sensory tracts or in the cerebral cortex.
B. Bilateral sensory loss in the hands or feet in a glove or stocking distribution.
Ans.   This is typical of a *peripheral neuropathy*, a condition commonly of metabolic etiology (e.g., diabetes or alcoholism), resulting in selective involvement of these regions. Symptoms are sometimes faked or hysterical.

C. Unilateral loss of all sensory modalities in an entire extremity.

Ans.　Unless the lesion encompassed all the nerve roots to that extremity, which is unlikely, the lesion is somewhere above the level of the spinal cord, in the contralateral brain stem, thalamus, or higher. A lesion below the brain stem, in the spinal cord, would not affect *all* sensory modalities, as only some of the nerve pathways have crossed over at spinal cord levels.

D. Loss of pain-temperature sense on the left side of the body below the neck, and paralysis and loss of proprioception-stereognosis below the neck on the right.

Ans.　Hemisection of the right cervical cord (the Brown-Sequard syndrome). All symptoms are ipsilateral in the Brown-Sequard syndrome, except for the contralateral pain-temperature loss below the level of the lesion.

3-6　A patient with a tumor experiences loss of pain-temperature in the left lower extremity followed by spastic paralysis on the right. Where is the tumor located?

　　　Ans.　In the right anterolateral aspect of the cord, compressing first the right spinothalamic tract and then enlarging to involve the right corticospinal tract.

Addendum: Information for touch, proprioception, stereognosis, and vibration also appears to travel in a pathway called the *spinocervicothalamic* tract (Fig. 18A). It is possible that the case problems in this book that are ascribed to "posterior column" lesions in reality involve the spinocervicothalamic tract as well. The simplest way of viewing this is to imagine that the dorsal columns extend laterally to include the *spinocervicothalamic* tract

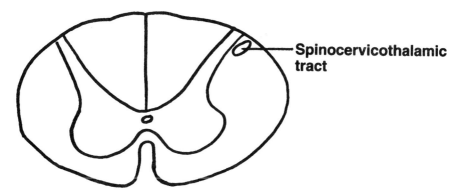

Spinocervicothalamic tract

Fig. 18A　The spinocervicothalamic tract.

CHAPTER 4.   BRAIN STEM

In evaluating a patient, the neurologist asks a sequence of questions. First, *where* is the lesion (spinal cord, brain stem, cerebrum, etc.)? Second, *what* is the lesion (tumor, infection, hemorrhage, etc.)? Third, what can be done to help the patient (medication, surgery, etc.)? The neurologist tries to determine if a single lesion can account for the patient's symptoms and signs. If multiple lesions must be postulated, this generally implies either metastatic disease, multiple sclerosis, the presence of two different diseases, or the presence of malingering or hysteria.

Precise anatomical localization of the lesion is very important in neurological diagnosis. It therefore is important to know the locations of the major fiber tracts and nuclei. The brain stem is very important in this regard. The following chapter describes the anatomy of the major nuclei and fiber pathways in the brain stem. Clinical illustrations of the usefulness of this data will be given in the questions following Chapter 5.

Don't panic. The brain stem is incredibly simple when studied in a logical sequence, as follows.

1.   Memorize the 12 cranial nerves and their functions. Ribald mnemonics will not help. You must know the individual cranial nerves and their functions on instant recall.

CN1:	Smells
CN2:	Sees
CNs 3, 4 and 6:	Moves eyes; CN3 constricts pupils, accommodates
CN5:	Chews and feels front of head
CN7:	Moves the face, tastes, salivate, cries
CN8:	Hears, regulates balance
CN9:	Tastes, salivates, swallows, monitors carotid body and sinus
CN10:	Tastes, swallows, lifts palate, talks, communication to and from thoraco-abdominal viscera
CN11:	Turns head, lifts shoulders
CN12:	Moves tongue

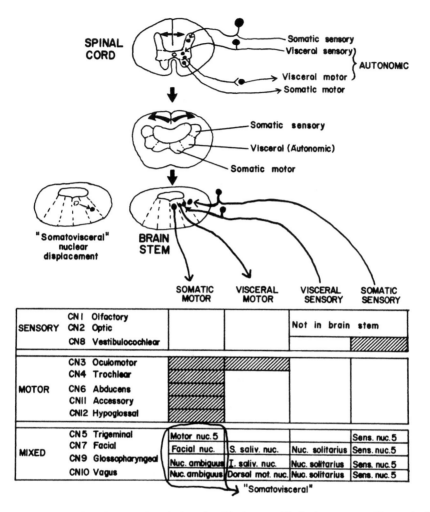

SPINAL CORD
— Somatic sensory
— Visceral sensory) AUTONOMIC
— Visceral motor
— Somatic motor

— Somatic sensory
— Visceral (Autonomic)
— Somatic motor

"Somatovisceral" nuclear displacement   BRAIN STEM

			SOMATIC MOTOR	VISCERAL MOTOR	VISCERAL SENSORY	SOMATIC SENSORY
SENSORY	CN1	Olfactory			Not in brain stem	
	CN2	Optic				
	CN8	Vestibulocochlear				///////
MOTOR	CN3	Oculomotor	///////	///////		
	CN4	Trochlear	///////			
	CN6	Abducens	///////			
	CN11	Accessory	///////			
	CN12	Hypoglossal	///////			
MIXED	CN5	Trigeminal	Motor nuc.5			Sens. nuc.5
	CN7	Facial	Facial nuc.	S. saliv. nuc.	Nuc. solitarius	Sens. nuc.5
	CN9	Glossopharyngeal	Nuc. ambiguus	I. saliv. nuc.	Nuc. solitarius	Sens. nuc.5
	CN10	Vagus	Nuc. ambiguus	Dorsal mot. nuc.	Nuc. solitarius	Sens. nuc.5

"Somatovisceral"

Fig. 19   The homology between the spinal cord and brain stem. The brain stem resembles a spinal cord stretched in the direction indicated by the midline arrows. The somatic and visceral components of the grey matter remain in the same relative position in spinal cord and brain stem. Chart indicates the major nuclei of the cranial nerves.

Note: a) There are 3 sensory nerves. b) There are 5 motor nerves of which one (CN3) has a visceral motor component in addition to a somatic motor component. The remainder of the motor nerves are purely somatic motor. c) There are 4 mixed (sensory plus motor) nerves. All of these have 4 components except for CN5, which has two components. d) Solitarius is Sensory (visceral) CN7, 9, 10. e) aMbiguus is Motor—CN9,10. f) CNs 5, 7, 9 and 10 are "somatovisceral"—see text for explanation. Nuc, nucleus; S., superior; I., inferior; Saliv., salivatory; Sens., sensory.

2.   Memorizing the cranial nerve functions will be easy if you memorize the chart in figure 19, which includes the same information as above rearranged differently. Unlike the spinal nerves, which are mixed nerves containing motor and sensory components, the cranial nerves are much simpler. Three of them are purely

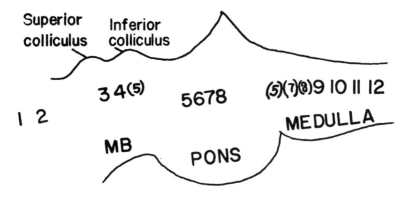

Fig. 20   The location of the cranial nerves and their nuclei within the midbrain (MB), pons, and medulla as seen laterally.

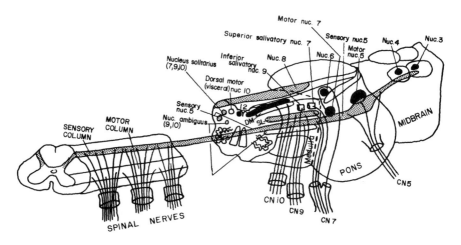

Fig. 21   A composite picture of the cranial nerve nuclei.
Note: a) The long motor and sensory columns of the spinal cord break up into individual brain stem nuclei. This occurs *gradually* (elongated nuclei in medulla, rounder in pons and midbrain). b) The sensory nucleus of CN5 does not break up but is continuous from spinal cord through midbrain. c) The motor nucleus of CN5 is confined to the pons. d) There are three sensory nuclei in the brain stem —the vestibulocochlear complex (CN8), sensory nucleus of CN5, and nucleus solitarius. e) Solitarius is a visceral sensory nucleus (CNs 7, 9, 10). f) The sensory nucleus of CN5 is somatic sensory (CNs 5, 7, 9, 10 and possibly others—but the predominant input is from CN5). g) a*M*biguus is *M*otor (CNs 9, 10). h) Nucleus solitarius is solitarily confined to the medulla as are nucleus ambiguus, the nucleus of CN12, the dorsal motor nucleus of CN10 and the salivatory nuclei. i) Nuc. 8 overlaps the medulla and pons. DM, Dorsal motor (visceral) nuc. 10; SL, Nucleus solitarius. (Modified from Dunkerley, G.B., *A Basic Atlas of the Human Nervous System,* F.A. Davis Company, 1975.)

sensory, 5 are purely motor and only four are mixed. The term "somatovisceral" will be explained later.

3.   In both spinal cord and brain stem the grey matter (nuclei) lies close to the central canal and the white matter (long fiber tracts) farther away. Also, the somatic and visceral motor and sensory components occupy the same relative

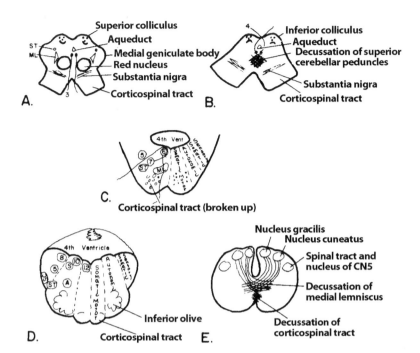

Fig. 22   How to draw cross sections through the brain stem. A. Rostral midbrain. B. Caudal midbrain. C. Pons. D. Rostral medulla. E. Caudal medulla. A, nucleus ambiguus; ML, medial lemniscus; S, nucleus solitarius; SC, spinocerebellar tract; ST, spino-thalamic tract; 3, 4, CNs 3 and 4; 5, nucleus and tract CN5; 6, 7, 8, nuclei CNs 6, 7, 8; 10, dorsal motor nucleus CN10; 12, nucleus CN12. Compare with figure 25.

positions. The cross sectional structure of the brain stem resembles that of a spinal cord that was grabbed posteriorly and stretched laterally (Fig. 19).

4.   The 12 cranial nerves and their nuclei are apportioned equally between the 3 brain stem segments (CNs 1, 2, 3, 4 in midbrain; CNs 5, 6, 7, 8 in pons; CNs 9, 10, 11, 12 in medulla). Actually, that's not quite true. CNs 1 and 2 lie not in the midbrain but more rostrally near the diencephalon. Also, although all divisions of CN5 enter the pons, the sensory nucleus of CN5 extends from midbrain to spinal cord (Fig. 27); the nuclei of CN8 lie not only in the pons but also in the medulla (Fig. 21); the nuclei and nerve entry points of CN7 are located in both pons and medulla (Fig. 28). These exceptions for CNs 5, 7, and 8 are indicated in brackets in figure 20.

5.   In general the cranial nerves exit relatively anterior to the ventricular system and do not cross over to the opposite side on exiting from the brain stem (Fig. 19). The only clinically important exception to this is dumb CN4, which both crosses over and passes over the roof of the brain stem (Fig. 23). No wonder the Lord delegated the most insignificant of all the cranial nerve functions to this untrustworthy nerve, that of moving the superior oblique muscle.

6. Figure 21 illustrates a basic difference between the spinal cord and brain stem. The central grey of the spinal cord is a long continuous column, in fact the longest "nucleus" in the CNS. Nerve fibers come off continuously from this large column and combine regularly to form the individual spinal peripheral nerves. As this grey column reaches the brain stem, it remains near the central canal but gradually becomes segmented into isolated nuclei. The segmentation is less pronounced in the medulla, where the nuclei remain elongated, and then becomes more evident in the pons and midbrain, where the nuclei are smaller and rounder.

7. Cranial nerve autonomic fibers are all parasympathetic (CNs 3, 7, 9, and 10).

How to Draw the Brain Stem in Cross Section

The rostral midbrain (Fig. 22A) resembles a double-headed gingerbread man with two belly buttons. In the caudal midbrain (Fig. 22B), the gingerbread man has lost his arms and has only one belly button.

The pons (Fig. 22C) resembles buttocks.

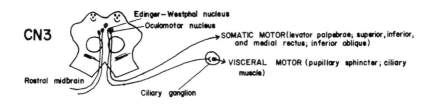

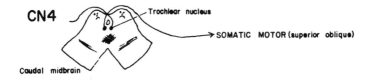

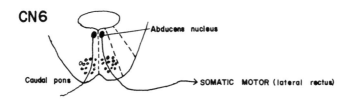

Fig. 23   Cranial nerves 3, 4, and 6.

The rostral medulla (Fig. 22D) has 3 bumps posterolaterally and 3 bumps anterolaterally—very convenient, for dotted lines between the bumps divide the medulla into the somatic motor, autonomic, and somatic sensory zones. The caudal medulla (Fig. 22E) looks something like the spinal cord with which it gradually merges.

With the above information, it becomes easy to predict and draw the anatomy of the cranial nerves with few exceptions, as follows.

The Five Motor Nerves (Figs. 23, 24)

The nuclei of CN3 are located, appropriately, near the aqueduct. Their axons, appropriately, exit anterolaterally without crossing over. Inappropriately, however, the *Edinger-Westphal* nucleus, which subserves visceral motor function, lies closer to the midline than predicted. Also, certain of CN3 fibers do cross over, but this need not concern us clinically.

CN4 is the exception to the rule of non-crossing over and anterior exit. It exists posteriorly, crossing the midline.

CN6 follows the rules. Its nucleus is located in the pons near the fourth ventricle in the somatic motor area, and it sends its fibers anteriorly without crossing over.

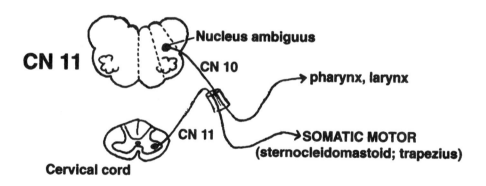

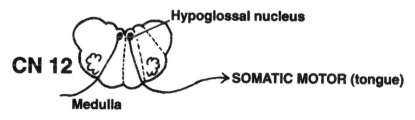

Fig. 24   Cranial nerves 11 and 12.

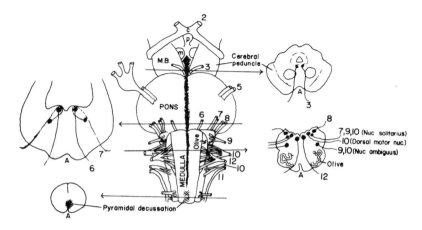

Fig. 25   Basilar (anterior) view of the brain stem. The outlines of four representative cross sections are shown at the levels of the horizontal lines. Structures that are externally visible in the intact brain stem are labelled in the cross sections. MB, midbrain, c, optic chiasm; p, pituitary gland; m, mammillary bodies; A, anterior; nuc., nucleus. Numbers refer to the individual cranial nerves.

CN11 is a social climber and really is not a cranial nerve. It is a spinal nerve which creeps through the foramen magnum, touches the vagus and comes back down again via the jugular foramen.

CN12 follows the rules perfectly. Note its exit anterior to the *inferior olivary nucleus* (a structure functioning in communication between cerebellum and other CNS areas). All the other cranial nerves in the medulla exit posterior to the olive (i.e., CNs 7, 8, 9, 10—see Fig. 25).

The Three Sensory Nerves

The nucleus of CN8 lies in its appropriate area (Fig. 26). After synapsing, the auditory fibers partly cross and travel to both inferior colliculi (Fig. 22), synapse again, extend to the medial geniculate bodies (Fig. 22), synapse, and then travel to both cerebral hemispheres (auditory area, Fig. 52). CNs 1 and 2, not belonging to the brain stem, will be discussed later.

The Four Mixed Nerves (5, 7, 9, 10)

If you have memorized figure 19, the four mixed nerves will then become predictable and easy, after the following story about the term "somatovisceral."

Once upon a time the brain stem had a dilemma. It knew the position occupied by the visceral nuclei, but was undecided on the definition of the term "visceral." It knew that "visceral" applied to smooth muscle, cardiac muscle and glands. However, in the vernacular, visceral also refers to the "gut," the digestive tract—which makes sense, for the stomach and intestines are smooth muscles. But what about the pharynx, for instance, which contains somatic striated muscle yet is part of the

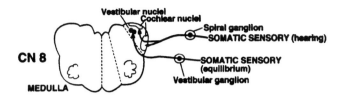

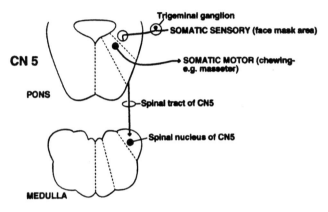

Fig. 26   Cranial nerves 8 and 5.

digestive tract? Is the pharynx visceral or somatic motor? If visceral, the pharyngeal nucleus would belong in the visceral segment of the brain stem. If somatic motor it would lie in the somatic motor portion. Undecided, the brain stem placed the pharyngeal nuclei and other such "somatovisceral" (also called "branchiomeric") nuclei within the visceral area but in a displaced position (Fig. 19) thus marking them as different. Each of the four mixed nerves has some somatovisceral motor component related to the gastrointestinal tract (5 - chewing; 7 - moving the lips; 9 and 10 - swallowing). Although some of their function are not related to the GI tract (e.g., CN7 - closing the eyes), the brain stem nevertheless lumped them all into the somatovisceral motor category. CNs 5, 7, 9 and 10, while technically called "somatic motor," contain no anatomical components in the somatic motor area of the brain stem.

Cranial Nerve 5

As noted in Fig. 26, CN5 has a somatic motor (actually somatovisceral motor) and a somatic sensory component in the appropriate regions. Note: the motor nucleus of 5 is a small round structure confined to the pons. The sensory nucleus of 5 is very long, extending into the medulla and becoming continuous with the posterior horn of the spinal cord (Fig. 21). Hence the need for a spinal tract to carry fibers along the length of the nucleus (Fig. 27). You see, all the motor and sensory fibers of CN5 enter the brain stem at the level of the pons; some of the sensory fibers then travel into the medulla and even the spinal cord via the spinal tract of 5 to synapse in the long sensory nucleus of 5.

The sensory distribution of CN5 is that of a face mask (Fig. 27). The sensory 5 distribution includes not only the superficial skin but also deeper tissues within the mask area, as is known to anyone who has experienced pain after belching soda up his nose, or anaesthesia in the dentist's office.

The proprioceptive, light touch and pain fibers of the trigeminal nerve are distributed respectively in a rostral to caudal sequence (Fig. 27) in the brain stem. Thus, the *tractotomy*, an operation severing the spinal tract of 5 at the level of the medulla to relieve unremitting facial pain, as in the condition known as *trigeminal neuralgia*, results in loss of pain with little change in other sensory modalities. The three branches of the trigeminal nerve are represented to some degree throughout the length of the sensory nucleus.

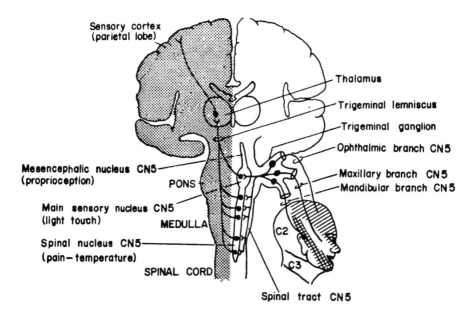

Fig. 27   Central sensory connections of the trigeminal nerve. The spinal nucleus of CN5 extends from the pons to spinal level C2. (Modified from DeMyer, W., *Technique of the Neurologic Examination*, McGraw-Hill Book Company, 1974.)

Cranial Nerve 7

CN7, the facial nerve, has four components (Fig. 28).

1.   The somatic (somatovisceral) motor component. This naughty nerve heard there was some sex to be had over by the 4th ventricle and began in early

development to grow in reverse direction, *into* the brain stem. On arriving at the 4th ventricle it realized that it misunderstood. It was six, not sex, and *the facial nerve* did an *about face*, traveling around the nucleus of 6, to exit anterolaterally. This component of the facial nerve is the most important and lies in the pons. The other three components lie in the medulla, as follows:

2.   The visceral motor component of CN7. The *superior salivatory nucleus*. The facial nerve, master of facial expression, cries when it is sad and spits when it is angry (Fig. 28). It lacrimates and salivates.

3.   The somatic sensory component of CN7 is a very minor contribution, monitoring sensation around the external ear. Indeed the somatic sensory components of CNs 7, 9, and 10 are all confined to the external ear, for the rest of the face has been hogged by CN5 and cervical nerve 2, the latter covering the area behind the head (Fig. 27). There is no sensory component to cervical nerve 1. Note that the sensory nucleus of 5 actually obtains somatic sensory input from cranial nerves, 7, 9, and 10 and possibly others (proprioception from CNs, 3, 4, 6), but it is still called the nucleus of 5 because of the overwhelming predominance of CN5 input.

4.   The visceral sensory component of CN7 (taste in the anterior two-thirds of the tongue) is also located precisely where it should be—in the visceral sensory area of the brain stem. Do not be upset by the fancy name of "nucleus solitarius" used to describe this visceral sensory nucleus. To understand why nucleus soli-

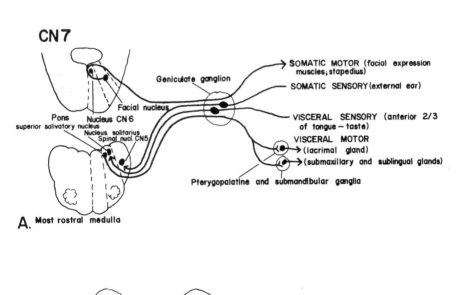

Fig. 28 A.   Cranial nerve 7. B. Visceral motor functions of the facial nerve.

33

tarius is not simply called the visceral sensory nucleus of 7 look back at figure 21. See how the grey matter columns of the spinal cord only *gradually* become broken up on passing through the brain stem. This makes the nuclei elongated in the medulla, meaning that a given nucleus may be shared by several cranial nerves at once. So nucleus solitarius cannot be called the visceral sensory nucleus of 7 when it is shared by CNs 7, 9, and 10. The same argument goes for *nucleus ambiguus* described below, which is shared by CNs 9 and 10.

The locations of nucleus solitarius and nucleus ambiguus give away their functions. Nucleus solitarius, unlike the very long *somatic* sensory nucleus of 5, is a *visceral* sensory nucleus located solitarily in the medulla in the visceral sensory area. Its visceral sensory modalities include taste (CNs 7, 9, 10), sensory input from the carotid sinus and carotid body (CN9) and the massive sensory return along the vagus nerve.

Nucleus ambiguus is located in the "somatovisceral" motor area of the medulla, coinciding with its somatovisceral function, namely swallowing (CNs 9 and 10) and speech (CN10).

Nucleus ambiguus is ambiguous for it is difficult to distinguish CNs 9 and 10 on examination; they both participate in the gag reflex and are usually described together rather than separately on exam. Of course, you could try to distinguish 9 and 10 by testing taste on the posterior tongue or epiglottis, but then you would run the grave risk of getting vomited upon. Speech, by the way, is not always noticeably affected with CN10 lesions despite unilateral vocal cord paralysis. One has to look into the throat to be sure.

Remember—Solitarius is Sensory 7, 9, 10; a*M*biguus is *M*otor 9, 10.

Nucleus solitarius and ambiguus become particularly important when evaluating the patient who presents with vertigo, a common clinical complaint. Vertigo (a sense of spinning) has two main causes. Sometimes it reflects dysfunction of the vestibular apparatus of the inner ear. At other times it results from dysfunction of the vestibular nuclei in the brain stem. Nucleus solitarius and ambiguus lie near the vestibular nuclei. Therefore, if the patient not only has vertigo, but also difficulty with taste, swallowing, or speech, this increases the suspicion of a brain stem lesion (see question 5-3c, syndrome of the posterior inferior cerebellar artery).

Cranial Nerve 9

Note in figure 29 the appropriate positions of the nuclei of CN9, as well as the functions of this nerve. CNs 9 and 10 each have their own superior and inferior ganglia.

Cranial Nerve 10

The nuclear arrangement of CN10 also fits appropriately (Fig. 29). CN10 has many functions. Whereas CNs 9 and 10 mediate swallowing, only CN10 controls

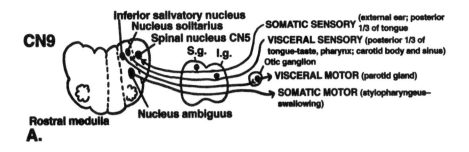

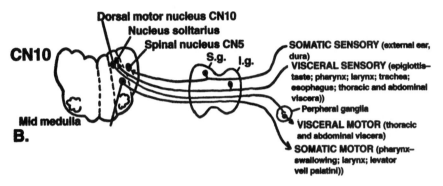

Fig. 29 Cranial nerves 9 and 10. S.g., superior ganglion; I.g., inferior ganglion.

the larynx. Moreover, it is a "vagrant" nerve traveling far to innervate the lungs, heart and abdominal viscera up to the splenic flexure of the large colon. Note the two motor nuclei of CN10—the autonomic one for the numerous autonomic functions of the nerve and the nucleus ambiguus for the somatovisceral functions of swallowing and innervation of the larynx.

CENTRAL CONNECTIONS OF CRANIAL NERVES 5 AND 7

Cranial Nerve 5

Just as all of the conscious sensory components of the spinal nerves eventually cross over and tend to synapse just prior to crossing, the same holds true for the trigeminal nerve. Sensory fibers enter the brain stem at the level of the pons and then, via the tract of 5, synapse ipsilaterally along the length of the trigeminal nucleus (Fig. 27). Then they cross over (with some minor exception not to be discussed) to form the trigeminal lemniscus. A lesion destroying one cerebral hemisphere would interfere with sensation on the opposite side of the body including the face. A lesion destroying one half of the brain stem (theoretical) would interfere with conscious sensation below the head contralaterally by destroying the medial lemniscus and spinothalamic tracts. The same lesion would decrease both ipsilateral and contralateral facial sensation (affecting both the ipsilateral nucleus and tract of 5 and ipsilateral trigeminal lemniscus which had crossed over).

The trigeminal lemniscus and medial lemniscus lie close to one another and for practical purposes may be considered to course in the same areas.

Cranial Nerve 7. The Facial Nerve

CN7 (a hook—Fig. 30) pulls down, closing the eye, whereas CN3 (3 pillars —Fig. 30) opens the eye. This is a vital clinical point.

Figure 31 illustrates the clinical difference between upper motor neuron and lower motor neuron damage to CN7. If one severs CN7, which innervates one entire side of the face, including the eyelids and eyebrows, that entire side of the face becomes paralyzed. The forehead on the affected side appears ironed out. The eye will not close and there is flattening of the nasolabial fold. If the lesion is above the level of the nucleus of 7, i.e., an upper motor neuronal lesion, only the area of the face below the eyes is paralyzed because of the bilateral innervation of the upper face by the two cerebral hemispheres. Hence, the typical stroke patient with an upper MN lesion rarely needs a tarsorrhaphy (an operation to keep the eyelids closed) whereas the patient with *Bell's palsy* (a spontaneous peripheral nerve 7 palsy of unknown etiology) may, because of marked difficulty in closing the eye. In severe cases, this results in drying of the eye unless the lids are sutured together or medication applied topically. The usual site of injury in Bell's palsy is somewhere in the facial canal (which lies between the *internal acoustic meatus* and the *stylomastoid foramen*—Fig. 32), and may involve other branches of CN7 (to the stapedius muscle, lacrimal and salivary glands—Fig. 32). Since the stapedius

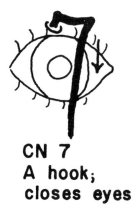

CN 7
A hook;
closes eyes

CN III
3 pillars;
opens eyes

Fig. 30   Eyelid innervation of cranial nerves 7 and 3.

dampens sound waves, its nonfunctioning leads to hyperacusis, wherein sounds appear excessively loud. Usually, function returns after Bell's palsy. The patient may then experience "crocodile tears," in which the patient tears on eating, instead of salivating, owing to misguided growth of the regenerating salivary and lacrimal fibers.

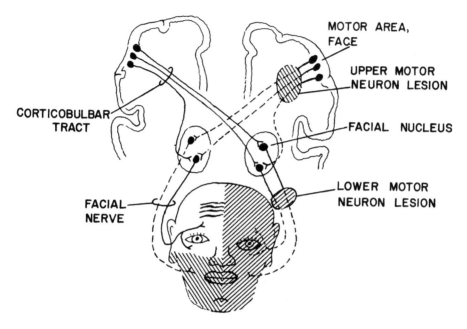

Fig. 31   Facial paralysis caused by upper and lower MN lesions of CN7. Shaded facial areas indicate zones of facial paralysis. (Modified from Clark, R.G., *Manter & Gatz's Essentials of Clinical Neuroanatomy and Neurophysiology,* F.A. Davis Co., Philadelphia, 1975.)

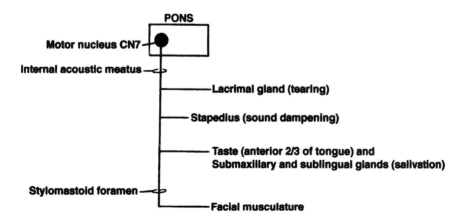

Fig. 32   Schematic view of the course of CN7.

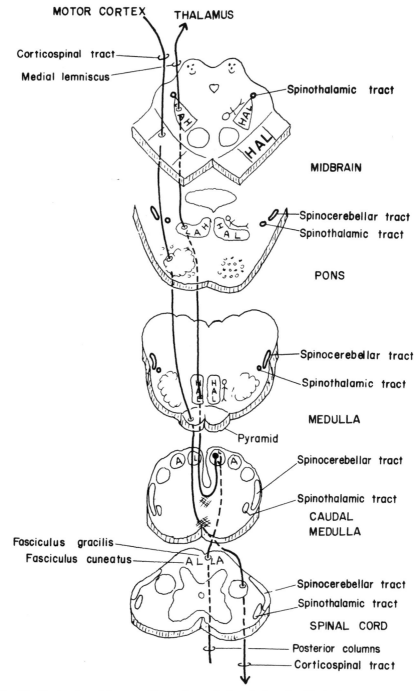

Fig. 33   The location of the major fiber pathways in the spinal cord and brain stem.

THE LONG TRACTS

The main pathways through the brain stem are illustrated in figure 33. The corticospinal tract, medial lemniscus, spinothalamic tract, and spinocerebellar tracts are the most important clinically. Not shown is the *corticobulbar tract*, which accompanies the corticospinal tract and connects, among other things, with the various brain stem motor nuclei. Unilateral damage to the corticobulbar tract does not cause massive dysfunction because each tract connects bilaterally with most cranial nerve motor nuclei. If one tract is damaged, the other tract carries on the function. The important exceptions are CNs 7 and 12, as corticobulbar tract lesions result in contralateral lower facial paralysis (Fig. 31) and, in some patients, contralateral weakness of the tongue (question 5-4). Bilateral lesions of the corticobulbar tracts result in a profound dysfunction, termed *pseudobulbar palsy*. Patients with this disorder experience difficulty not only with facial expression and motion of the tongue, but with chewing, swallowing, speech, and breathing. There may also be inappropriate spells of laughing or crying.

For practical clinical purposes, one may assume that corticobulbar fibers cross over contralaterally at about the same brain stem level as their motor nuclei of termination.

The topographical arrangement of HAL (Head, Arm, Leg) is better known for the medial lemniscus than the other fiber pathways. Picture Mr. HAL standing upright in the medial lemniscus of the medulla (Fig. 33). On approaching the sleep area in higher brain centers he gets sleepier and sleepier until he is lying down at the level of the midbrain. As you may be getting sleepy right now, this chapter is hereby ended.

Question

4-1. A patient experiences decreased conscious proprioception in the left upper extremity, decreased pain-temperature sensation in the left face, and decreased pain-temperature sensation on the right side of the body below the head. Where is the lesion?

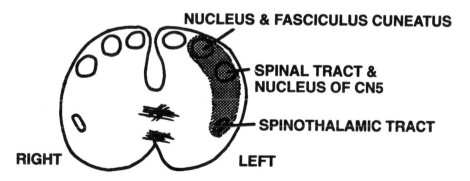

Fig. 33A   Ans. Caudal medulla (or upper cervical cord).

CHAPTER 5. THE VISUAL SYSTEM

Destruction of a cerebral hemisphere results in dense paralysis and sensory loss in the contralateral extremities. Such lesions do not result in corresponding visual loss or ocular paralysis confined to the contralateral eye. Rather, both eyes are affected partially. Neither eye can move to the contralateral side and neither eye sees the contralateral environment (Fig. 34).

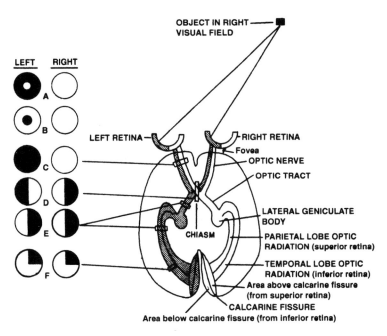

Fig. 34   The visual pathways as seen from above the brain. Letters A-F refer to visual field defects following lesions in the corresponding brain areas. Circles indicate what the left and right eyes see (the left and right visual fields). Black areas represent visual field defects. A. Constricted field left eye (e.g., end-stage glaucoma). When constricted fields are bilateral, it sometimes signifies hysteria. B. Central scotoma (e.g., optic neuritis in multiple sclerosis; macular degeneration), C. Total blindness of the left eye. D. Bitemporal hemianopia (e.g., pituitary gland tumor). E. Right homonymous hemianopia (e.g., stroke). F. Right superior quadrantanopia.

Optic fibers temporal to the fovea connect with the brain ipsilaterally. Fibers nasal to the fovea cross over to the opposite side at the optic chiasm. A lesion of the optic tract on the right, therefore, results in loss of the left visual field in each eye.

Note (Fig. 34) that the left visual field falls on the right half of each retina; the superior visual field falls on the inferior retina. The left visual field projects to the right side of the brain. Similarly, the superior visual field projects below the calcarine fissure in the occipital lobe. In other words everything is upside down and backwards, provided you think in terms of *visual fields*. For example, a patient with a lesion below the right *calcarine fissure* experiences a left superior field deficit (Fig. 35).

The center of the retina (the *fovea*, which is the area of most acute vision) projects to the tip of the occipital lobe (Fig. 35). Thus, a patient with a severe blow to the back of the head may experience bilateral central scotomas (visual field defects) if both occipital poles are destroyed.

Commonly, visual field defects partially spare the macula (the area of retina in the immediate vicinity of the fovea). This may be due to the large representation of the macula on the visual cortex, with overlapping blood supplies, resulting in only partial lesions. It may also be due to artifacts of fixation in the course of the exam, or to a certain degree of bilateral representation of the macula.

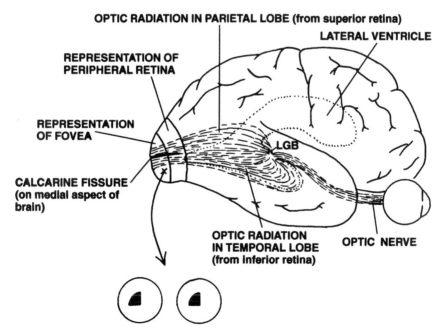

Fig. 35   Lateral view of the visual pathways, showing the visual field defect resulting from a lesion below the right calcarine fissure in the foveal region of the occipital cortex. LGB, lateral geniculate body. (Modified from Cushing, H., Trans. A. Neurol. Assoc. 47:374-433, 1921.)

OPTIC REFLEXES

Pupillary Constriction to Light

Unlike the pathways mediating vision, which involve a synapse in the lateral geniculate body, the pupillary light reflex involves a direct pathway to the midbrain from the optic tract (Fig. 36). Shining a light in one eye normally leads to constriction of both pupils (termed the *consensual reflex*) as may be deduced from the connections depicted in figure 36.

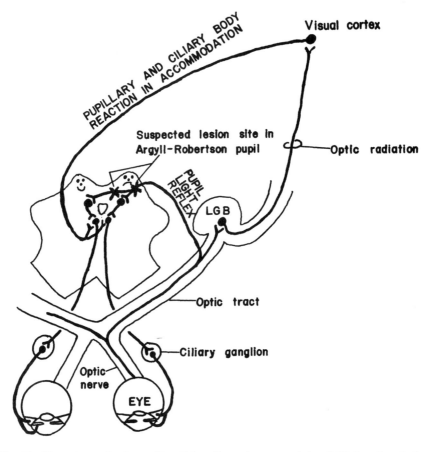

Fig. 36   The pathways for the pupillary light reflex and accommodation. LGB, lateral geniculate body. The depicted lesions presumably also interrupt light reflex fibers crossing from the opposite side of the brain stem. The pathway shown innervating the eye is highly schematic; the light reflex pathway involves only pupillary constriction, whereas the accommodation pathway affects both pupillary constriction and ciliary body accommodation.

Accommodation

Accommodation (Fig. 36) involves a neural circuit to the visual cortex and back, which makes sense, for we need our cerebral cortex to determine that something is out of focus before we can send directions to correct the focus. Focusing occurs by stimulating the smooth muscle of the ciliary body in the eye to contract, thereby enabling the lens to change its shape (accommodation). During accommodation not only does the lens focus but the pupil constricts, both smooth muscle actions mediated by parasympathetic components of CN3.

The syphilitic (*Argyll-Robertson*) pupil (also called the prostitute's pupil because it accommodates but does not react) constricts during accommodation, which is normal, but does not constrict to light. The lesion is considered to lie in the *pretectal area* of the superior colliculus (Fig. 36).

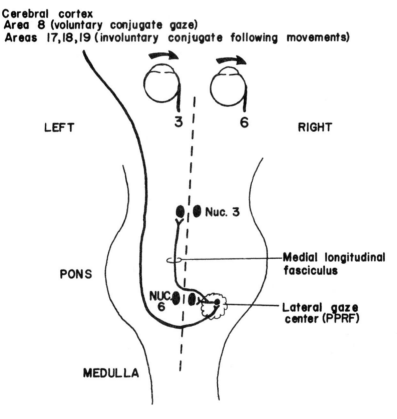

Fig. 37   The pathway for lateral conjugate gaze.
PPRF—pontine paramedian reticular formation.

Conjugate Gaze

Damage to the motor areas of the cerebral cortex produces contralateral paralysis of the extremities. It does not produce loss of all the contralateral eye muscle movements, but rather loss of the ability of either eye to look toward the contralateral environment. Following a lesion to the left visuomotor area (Brodmann's area 8—see Fig. 52), the patient cannot voluntarily look to the right. His eyes tend to deviate to the left. In essence, they "look at the lesion." This occurs because the pathway from the left hemisphere innervates the right lateral rectus muscle (right CN6) and the left medial rectus muscle (left CN3). See figure 37. The right lateral rectus and left medial rectus muscles both direct the eyes to the right.

A lesion to the *medial longitudinal fasciculus* bilaterally, most commonly seen in multiple sclerosis, would produce a decreased ability for either eye to look medially (Figs. 37, 38). In this instance, both eyes could converge because the pathway for convergence (as well as vertical gaze) is different from the path for lateral conjugate gaze. The condition resulting from lesions to the MLF is known as the *MLF (medial longitudinal fasciculus) syndrome, or internuclear ophthal-moplegia* (Fig. 38).

In the *one-and-a-half syndrome*, there is a lesion of one abducens nucleus, as well as the crossed connections to both MLFs. Thus, in a right-sided lesion, neither eye can look to the right on attempted right conjugate gaze. On attempted left conjugate gaze, only the left eye moves, as its lateral rectus function is preserved.

Convergence and vertical gaze apparently involve circuits in the midbrain close to (although not within) the superior colliculus. Hence, difficulty with convergence and vertical gaze may arise in tumors of the pineal gland which press upon the brain stem at the superior collicular level. *Parinaud's syndrome* is pupillary constriction and paralysis of vertical gaze following lesions close to the superior colliculus.

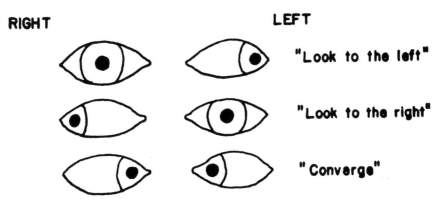

Fig. 38   MLF syndrome.

Nystagmus

Nystagmus is a repetitive, tremor-like oscillating movement of the eyes. The most common form of nystagmus is *horizontal jerk nystagmus*, wherein the eyes repetitively move slowly toward one side and then quickly back. It is normal to have a slight degree of such nystagmus on attempted extreme lateral gaze, but marked degrees are abnormal and found in a variety of clinical conditions. *Vertical nystagmus* is always abnormal, signifying a disorder in brain stem function. *Pendular nystagmus*, in which the eye moves at equal speeds in both directions, commonly is congenital or present after prolonged periods of blindness.

The pathways for nystagmus are so complicated that the situation is best explained by a little story.

It is common knowledge that the brain stem is very dumb and slow, being the most primitive part of the brain. The cerebral cortex, a relatively recent evolutionary development, is smart and fast. One day, a man was driving along in his car when someone sitting to his right squirted some cold water into his right ear (the clinical test known as *cold calorics*). The, dumb brain stem sluggishly moved its eyes to the right, drawling in its deep, slow tone "HEY-Y-Y-Y-Y! WHO'S SQUIRTING COLD WATER IN MY EAR?" Alarmed at this, the smart, quick cortex quickly jerked the eyes back, shouting "HEY! GET BACK AND LOOK AT THE ROAD!" These are the slow and fast components of nystagmus following cold calorics. The clinical usefulness of this test will soon become apparent, after the following bit of additional information.

We all know that as consciousness declines, say in sleep, the cerebral cortex becomes depressed more readily than the brain stem. This is fortunate, for if the brain stem went to sleep we might lose some rather vital functions like control of respiration, which is located in the brain stem. It takes a pretty severe

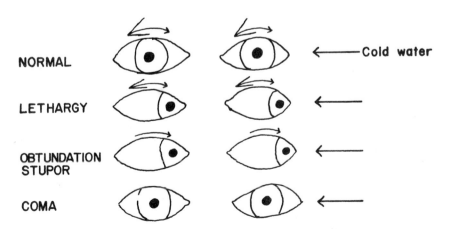

Fig. 39 Cold caloric responses in states of altered consciousness. Small arrow indicates the slow phase of nystagmus. Large arrow indicates the fast phase.

environmental change to depress the brain stem. With the above information we can now predict the reflex ocular responses to cold caloric testing and understand its clinical application.

State of Consciousness

In the normal waking state (Fig. 39), when cold water is introduced into the ear there is nystagmus, with the slow component (small arrow) toward the side of stimulation, followed by a fast component (large arrow) in the opposite direction. There is no net deviation of the eyes. As the patient becomes lethargic the cortex begins to fall asleep first so the fast component of nystagmus becomes less pronounced. There still is nystagmus but also a net deviation of the eyes. With further depression (obtundation, stupor) there is deviation without the fast movement of nystagmus, for the cerebral component is defective. With coma, the brain stem becomes depressed and there is no movement at all. Whereas the last pattern occurs in true coma, the first pattern appears in the patient faking coma.

Muscle Paresis

Cold calorics can confirm the suspicion of a CN3 or CN6 paresis in a stuporous patient (Fig. 40). For instance, the *uncus*, a portion of the temporal lobe lying particularly close to CN3 (Fig. 9) may herniate against CN3 and damage it following a subdural hemorrhage, resulting in a widely dilated, fixed (unresponsive to light) pupil. Cold calorics, by confirming the presence of medial rectus weakness, helps to confirm that the dilated pupil is due to CN3 injury and not to other causes (e.g., someone placing dilating drops in the eye and not telling anyone).

Doll's Eye Phenomenon

On turning the head suddenly to one side there is a tendency in patients without brain stem damage for the eyes to lag behind. This is the Doll's Eye phenomenon, as in a doll's eye which looks down when the head is tilted back. The reflex is believed to be brain stem mediated and any asymmetry or lack of response is believed to reflect significant brain stem dysfunction. Of course, it is wise to know, before performing this exam, that the patient does not have a broken neck.

Fig. 40   Cold caloric response in a case of right third nerve paralysis.

Questions

5-1   A 53-year-old woman reports that her nucleus ambiguus isn't feeling well. What are her symptoms?
Ans.   Hoarseness (often but not always present with CN10 lesions) and difficulty swallowing.

5-2   A patient chronically ingesting full doses of streptomycin, quinine, and aspirin began complaining of bilateral hearing loss. Why?
Ans.   All these agents can damage the auditory nerve.

5-3   Draw a single unilateral lesion accounting for the following symptoms and signs.
    A. Right hemiplegia, weakness of the left tongue (tongue deviates to left when protruded), atrophy and fasciculation of the left tongue.

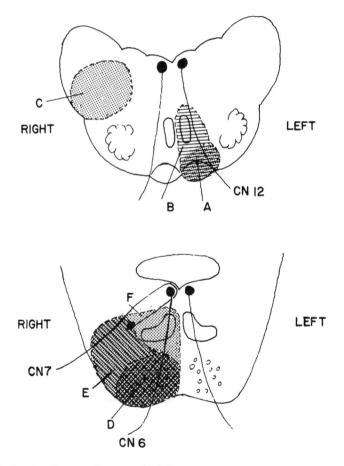

Fig. 41   Lesion sites. Letters refer to question 5-3.

Ans.   Medulla, left pyramid and left hypoglossal nerve (Fig. 41A). In riding a bicycle, pushing against the right handlebar will cause the wheel to turn to the left. Similarly, weakness of the left tongue results in deviation of the tongue to the left on attempted protrusion.

B. The above symptoms plus conscious proprioceptive loss in the right upper and lower extremities.

Ans.   Left pyramid, plus left hypoglossal nerve plus left medial lemniscus, as might occur in an occlusion of a branch of the anterior spinal artery (Fig. 41B).

Note in figure 9 that the crotch of Willis lies at the junction of the pons and medulla. This means that the anterior spinal artery, vertebral artery and posterior inferior cerebellar artery all run over the medulla for part of their courses. All three of these vessels give off branches to the medulla, the anterior spinal artery being most medial, supplying the area noted in the above lesion. The most lateral area of the three is supplied by the posterior inferior cerebellar artery (Fig. 42). The pons is supplied by vascular branches from the basilar artery, anterior inferior cerebellar artery, and superior cerebellar artery.

C. Cerebellar dysfunction with right sided ataxia, loss of pain-temperature on the right face and left upper and lower extremities, hoarseness, difficulty swallowing, loss of taste on the right, vertigo and nystagmus.

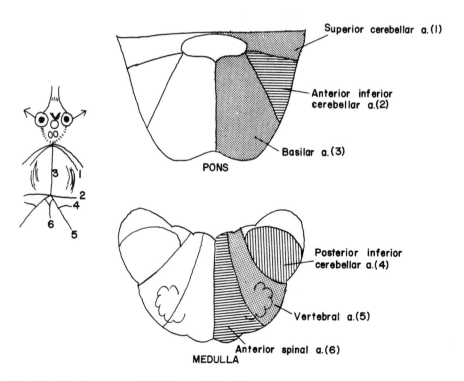

Fig. 42   Vascular supply to the brain stem (schematic).

Ans. This is the classical *syndrome of the posterior inferior cerebellar artery* (Fig. 41C) which may become thrombosed. Cerebellar dysfunction— right spinocerebellar tract; loss of right facial pain-temperature—injury to spinal tract and nucleus of right CN5. Loss of left upper and lower extremity pain-temperature—injury to right spinothalamic tract (remember, it crossed over); hoarseness and swallowing difficulty—right nucleus ambiguus; taste —right nucleus solitarius; nystagmus—irritation of the vestibular nuclei.

D. The right eye cannot abduct (move laterally); left hemiplegia.

Ans. Lesion to right pons involving CN6 plus right corticospinal tract (Fig. 41D).

E. Case D plus total right peripheral facial paralysis.

Ans. Pons including fibers from CN7 (Fig. 41E).

F. Case E plus loss of left position and vibratory sense.

Ans. Pons with involvement of right medial lemniscus (Fig. 4IF).

G. Progressive left-sided deafness and left lower motor neuron facial paralysis, absent left-sided cold caloric responses.

Ans. A defect in hearing involves either the cochlear nuclei, CN8, or more peripheral areas as there is little effect on hearing with lesions central to these levels (bilateral representation of hearing). The lesion may involve a tumor of the ponto-cerebellar angle (the angle between pons, medulla, and cerebellum), e.g. an acoustic neuroma, affecting CNs 7 and 8.

H. Paralysis of the left arm and leg and right masseter muscle. Right facial anesthesia (loss of all sensation); anesthesia of left upper and lower extremities, paralysis of left tongue, paralysis of left lower facial muscles, paralysis of conjugate gaze to the left.

Ans. A lesion in the pons will affect the right corticospinal tract, right motor nucleus of 5 (a small nucleus confined to the pons), right sensory nucleus of 5, right spinothalamic tract and medial lemniscus. There may also be interruption of the corticobulbar tract on the right before the fibers cross over to the left nuclei of CNs 12 and 7. Paralysis of left lateral gaze involves interruption of the conjugate gaze pathways prior to crossing over to the left lateral gaze center (Fig. 43H).

I. Paralysis of the right lower facial muscles and right upper extremity and inability to adduct (move toward the body midline) the left eye; left ptosis and dilation of the left pupil. Tongue deviates to the right side.

Ans. Midbrain lesion. Paralysis of right upper extremity—medial aspect of left pyramidal tract, including as yet uncrossed fibers (upper motor neuron) to right nuclei of CNs 7 and 12. Damage to fibers of left CN3 results in left ptosis and dilation of the left pupil (Fig. 43I).

J. Inability to adduct the left eye, with diplopia (double vision) on right lateral gaze plus right hemianesthesia to all sensory modalities.

Ans. Left CN3 involvement at the level of the midbrain (Fig. 43J); involvement of both the left medial lemniscus and left spinothalamic tracts which are in direct apposition at this level (in contrast to their separate positions at the level of the medulla—see Fig. 33).

K. Left-sided headache, total paralysis left side of face with vertigo and left-sided hearing loss; no other neurological deficit.

Ans.   Possible left-sided cerebellopontine angle tumor (outside the brain stem).

L. Nystagmus, bilateral internuclear ophthalmoplegia, central scotoma right eye, weakness right lower extremity with right Babinski, urinary incontinence, right ptosis with difficulty adducting right eye.

Ans.   Multiple lesions must be postulated, probably secondary to multiple sclerosis. The other major cause of multiple central nervous system lesions is metastatic cancer.

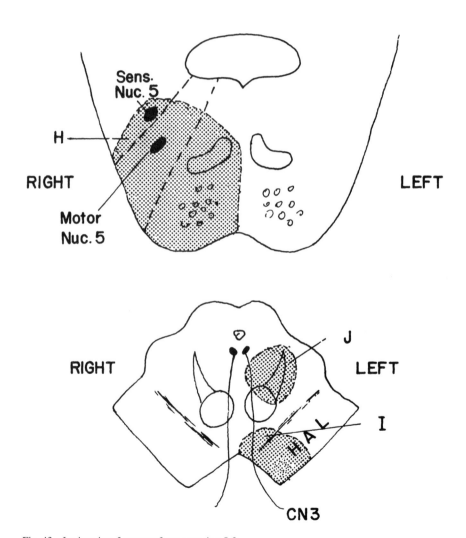

Fig. 43   Lesion sites. Letters refer to question 5-3.

5-4   What deficits will occur in the cranial nerves following destruction of the left cerebral hemisphere?
Ans.   CN1—loss of olfaction on the left. CN2—right homonymous hemianopia. CNs 3, 4, 6—loss of right conjugate gaze. CN5—right facial hemianesthesia. Only slight motor defect in chewing on the right, owing to bilateral innervation of the motor nucleus of 5. CN7—lower right facial paralysis. CN8—little defect owing to bilateral representation. CNs 9, 10—no significant defect—bilateral innervation. CN11—decreased function on the right (minimal). CN12—variable contralateral tongue paresis.

You can remember the motor defects with most CN upper MN lesions by noting which muscle maneuvers you find difficult to perform. Most of us will experience difficulty in raising one eyebrow, trying to move but one eye, biting down exclusively on one side, or swallowing or talking on one side only. These difficulties coincide with bilateral innervation of the cranial nerves.

5-5   Who comes to the doctor's office first—A patient with a constriction of the visual fields as in end stage glaucoma or a patient with a central scotoma, as in multiple sclerosis (Fig. 34)?
Ans.   The patient with a central scotoma. It is associated with a profound drop in visual acuity owing to foveal involvement. The patient with a constricted field may have 20-20 vision and be unaware of a gradually constricting field. Hence the reason for routine office glaucoma screening.

5-6   A one-eyed patient has the visual field depicted in figure 44. Where is the lesion?
Ans.   It is impossible to tell whether the lesion corresponds to a right homonymous hemianopia or to a bitemporal hemianopia. The lesion either involves the optic chiasm or is located somewhere between the chiasm and the visual cortex on the left.

5-7   A friendly patient came to the emergency room for a foot rash. The exam was normal except that the physician noted a widely dilated right pupil. He called the neurosurgeon to perform immediate emergency burr holes in the patient's skull to drain his subdural hematoma, which apparently must have caused an uncal herniation and CN3 damage. What did the physician fail to consider?

Fig. 44   Visual field defect in question 5-6.

Ans.   Patients with uncal herniation and brain stem compression tend to be quite sick. This patient wasn't. A dilated pupil may be caused by accidental instillation of dilating drops in the eye or congenitally as in *Adie's pupil*. In the latter, the pupil is always dilated and responds minimally and sluggishly to light. It is unaccompanied by other CN3 signs such as ptosis or oculomotor defects.

5-8   A patient's left eye looked funny. The left upper lid appeared higher than on the right. His pupil was larger on the left than on the right. What single lesion may account for this?
Ans.   *Horner's syndrome* on the right, interrupting the sympathetic fibers to the eye on the right. Note that it may be difficult by appearance alone to determine which side a defect is on. In this case the defect was on the right and the patient's appearance was misleading.

5-9   A patient has no pupillary reaction at all to light shined on the left side. There is a reaction to light in both eyes when light is shined on the right. Where is the lesion?
Ans.   Left CN2.

5-10   There is pupillary reaction to light on the right side only, when light is shined on the left or right eye. Where is the lesion?
Ans.   Left CN3 lesion.

5-11   Where is the lesion in a patient with inability to move the right eye past the midline on attempted left conjugate gaze but otherwise has good convergence ability?
Ans.   Right medial longitudinal fasciculus (internuclear ophthalmoplegia). See figure 37.

5-12   There is only one area of the nervous system where the nuclei of primary sensory neurons are located *within* the CNS rather than in outside ganglia. This is the *mesencephalic nucleus of CN5* (Fig. 27) which contains the nuclei of CNS proprioceptive fibers. Is this an important clinical fact?
Ans.   No.

5-13   In the corneal reflex, both eyes normally blink upon touching either cornea. If neither eye blinks on touching the right cornea and both eyes blink on touching the left cornea, which cranial nerve is likely to be affected?
Ans.   Right CN5. CN5 carries corneal sensation, whereas CN7 innervates the orbicularis oculi muscle.

5-14   If only the left eye blinks on touching either the right or left cornea, which cranial nerve is likely to be involved?
Ans.   Right CN7.

5-15 If you stick a finger in your left ear and touch a tuning fork to the forehead midline (or simply hum), which ear hears the sound loudest?
Ans. The left ear, for you have simulated a conductive loss on the left, in which bone conduction is accentuated. Ear wax occlusion and damage to the tympanic membrane or ossicles also cause conductive losses as opposed to neural hearing loss which results from CN8 injury.

5-16 Why should one test CN1 in a patient with a suspected optic chiasm lesion?
Ans. The olfactory tracts (Figs. 9, 47) pass close to the optic chiasm.

5-17 Describe the actions of the superior and inferior rectus muscles and superior and inferior oblique muscles with regard to elevation and depression of the eye.
Ans. The superior rectus and inferior oblique muscles both elevate the eye. The inferior rectus and superior oblique muscles depress the eye. The rectus muscles act maximally when the eye is deviated temporally. The obliques act maximally when the eye is deviated nasally.

5-18 Bilateral lesions of the cerebrum or ventricular regions of the brain stem may result in abnormal breathing patterns. The precise neuronal circuitry involved is unclear. Generally localize the lesion in the following breathing patterns.
Ans.

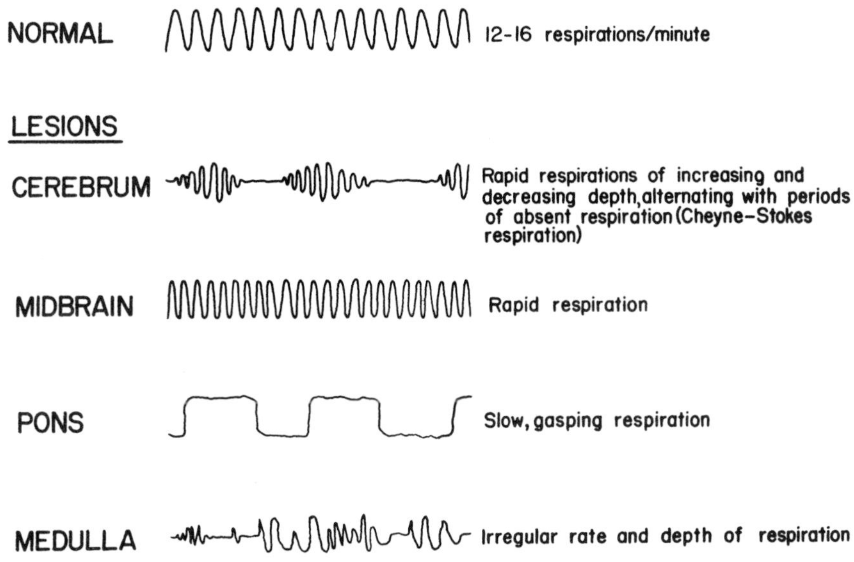

NORMAL 12-16 respirations/minute

LESIONS

CEREBRUM Rapid respirations of increasing and decreasing depth, alternating with periods of absent respiration (Cheyne-Stokes respiration)

MIDBRAIN Rapid respiration

PONS Slow, gasping respiration

MEDULLA Irregular rate and depth of respiration

Fig. 44A

CHAPTER 6.   AUTONOMIC SYSTEM AND HYPOTHALAMUS

The autonomic system regulates glands, smooth muscle and cardiac muscle. It contains sympathetic and parasympathetic components. The sympathetic system as a whole is a catabolic system, expending energy, as in the flight or fight response to danger, e.g. increasing the heart rate and contractility and shunting blood to the muscles and heart. The parasympathetic system is an anabolic system,

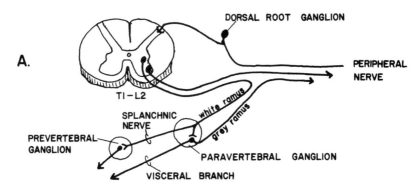

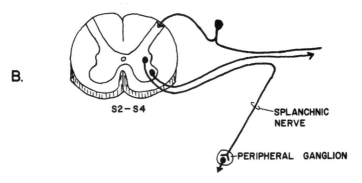

Fig. 45 A.   The sympathetic nerve routes.   B.   The parasympathetic nerve routes.

conserving energy, e.g. in slowing the heart rate and in promoting the digestion and absorption of food. The cell bodies of preganglionic sympathetic fibers lie in the *intermediolateral columns* of the spinal cord at spinal cord levels T1-L2 (Figs. 45, 46). Those of the parasympathetic system occupy comparable positions at spinal cord levels S2-S4.

In addition, cranial nerves 3, 7, 9 and 10 have parasympathetic components (3—pupil and ciliary body constriction; 7—tearing and salivation; 9—salivation; 10—the vagus and its ramifications).

In order to extend all over the body the sympathetic system fibers leave the spinal cord at levels T1-L2, enter the paravertebral ganglion chain and then may travel up or down the chain for considerable distances prior to synapsing (Fig. 46). The sympathetic chain stretches from the foramen magnum to the coccyx and supplies the far reaches of the body with post-ganglionic sympathetic fibers. Parasympathetic fibers reach widespread areas via the vagus (Fig. 46).

Both parasympathetic and sympathetic systems contain two neurons between the spinal cord and periphery. The first synapse is cholinergic (containing acetylcholine). For the sympathetic system this synapse is either in the *paravertebral* chain of sympathetic ganglia or farther away in the *prevertebral* ganglion plexuses (Fig. 45A).

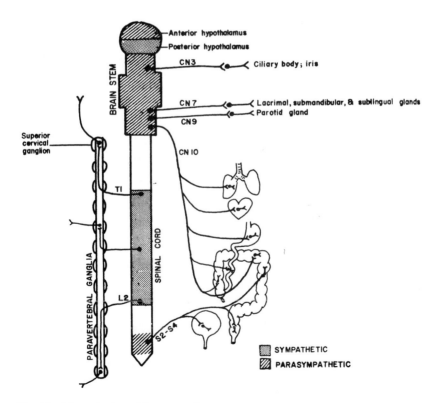

Fig. 46   Schematic view of the autonomic nervous system.

Parasympathetic synapses typically lie very close to or within the viscera. The final synapse of the parasympathetic system contains acetylcholine, whereas the final synapse of the sympathetic system contains noradrenaline, with the exception of certain synapses, as for sweating, that contain acetylcholine (i.e., are *cholinergic*). In the chart below, note that secretory functions in general are stimulated by cholinergic fibers.

Structure	Sympathetic function	Parasympathetic function
Eye	Dilates pupil (mydriasis) No significant effect on ciliary muscle	Contracts pupil (miosis) Contracts ciliary muscle (accommodation)
Lacrimal gland	No significant effect	Stimulates secretion
Salivary glands	Stimulates viscous secretion	Stimulates watery secretion
Sweat glands	Stimulates secretion (cholinergic fibers)	No significant effect
Heart		
Rate	Increases	Decreases
Force of ventricular contraction	Increases	Decreases
Blood vessels	Dilates or constricts cardiac & skeletal muscle vessels* Constricts skin and digestive system blood vessels	No significant effect
Lungs	Dilates bronchial tubes	Constricts bronchial tubes Stimulates bronchial gland secretion
Gastrointestinal tract	Inhibits motility and secretion	Stimulates motility and secretion
GI sphincters	Contracts	Relaxes
Adrenal medulla	Stimulates secretion of adrenaline (cholinergic fibers)	No significant effect
Urinary bladder	Relaxes (minimal)	Contracts
Sex organs	Ejaculation	Erection**
Liver	Stimulates gluconeogenesis & glycogenesis	
Fat	Stimulates lipolysis	
Kidney	Stimulates renin release	

*Stimulation of beta-2 receptors *dilates* cardiac and skeletal muscle vessels whereas stimulation of alpha-1 receptors *constricts*. Most dilation of cardiac and skeletal vessels, though, may be due to nonautonomic, local tissue *autoregulatory* responses to lack of oxygen.
**Parasympathetic = erection, since parasympathetic is a longer word.

In extreme fear both systems may act simultaneously, producing involuntary emptying of the bladder and rectum (parasympathetic) along with a generalized sympathetic response. In more pleasant circumstances, namely in sexual arousal, the parasympathetic system mediates penile and clitoral erection and the sympathetic controls ejaculation.

Proceeding rostrally from the caudal tip of the spinal cord, one first finds a parasympathetic area (S2-S4), followed by a sympathetic region (T1-L2), then parasympathetic areas (CNs 10, 9, 7, 3) and then successively a sympathetic and parasympathetic area, a strange alternating sequence. The latter two areas are the posterior and anterior parts of the hypothalamus (Fig. 46).

The hypothalamus, a structure about the size of a thumbnail, is the master control for the autonomic system. Stimulation or lesions result not in isolated smooth muscle, cardiac muscle or glandular effects but in organized actions involving these systems, e.g. in the fear or rage reaction of the flight or fight response, in increased and decreased appetite, altered sexual functioning, and control of body temperature. For instance, stimulation of the posterior hypothalamus may result in conservation of body heat and an increase in body temperature owing to constriction of cutaneous blood vessels.

Many circuits connect the hypothalamus with various areas of the cerebral cortex, brain stem and thalamus. Figure 47 shows the reverberating *Papez circuit* believed to be involved in the emotional content of conscious thought processes and in memory. It provides intercommunication between hippocampus, hypothalamus, thalamus and cerebral cortex. Note the input of the olfactory system, which also plays a role in emotion. This is evident if you have ever seen two

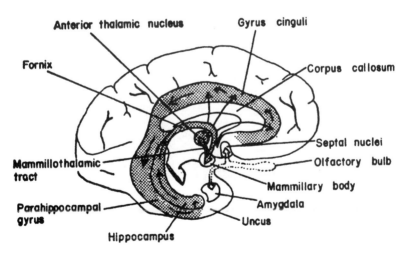

Fig. 47   The Papez circuit (shaded areas). The labeled structures as a whole are referred to as the *limbic system*. The hippocampus, among other things, is involved in the storage of short-term memory. (Modified from Clark, R.G., *Manter and Gatz's Essentials of Clinical Neuroanatomy & Neurophysiology*, F.A. Davis Company, Philadelphia, 1975.)

dogs sniffing one another or noted the prominence of the perfume industry in major department stores. In the *Klüver-Bucy syndrome*, there are lesions of the amygdala, resulting in docile behavior, hypersexuality, compulsive attentiveness to detail, and, with injury to nearby areas, visual agnosia (inability to recognize objects visually).

In *Wernicke's syndrome*, which occurs in patients who are alcoholic and under-nourished, there is paralysis of eye movements, ataxic gait and disturbances in the state of consciousness associated with hemorrhages in the hypothalamus and other regions.

Korsakoff's syndrome also occurs in alcoholic patients and consists of memory loss, confusion and confabulation associated with lesions in the mammillary bodies and associated areas.

The hypothalamus lies close to the pituitary gland. A disorder of one structure may affect the other. Figure 48 shows the major hypothalamopituitary connections. Note that nerve fibers from the *paraventricular* and *supraoptic nuclei* connect with the posterior pituitary. These nuclei secrete the hormones *oxytocin* and *antidiuretic hormone (vasopressin)*. These hormones are synthesized and transported in neurons and then released at the ends of the nerve terminals in the posterior pituitary.

The anterior pituitary contains no neuronal connections. Instead, *releasing factors* are produced in the hypothalamus and are released into the portal circulation and then transported to the anterior pituitary where they stimulate cells in the anterior pituitary to secrete various hormones, including adrenocorticotrophic hormone, thyrotrophic hormone, somatotrophic hormone, follicle stimulating hormone and luteinizing hormone.

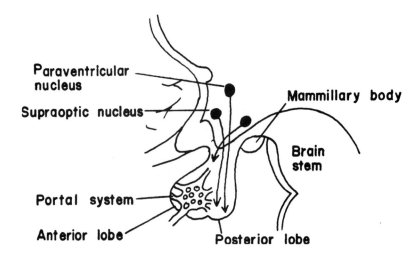

Fig. 48   The connections between the hypothalamus and pituitary gland.

Disorders of the autonomic system include:

1. *Riley-Day syndrome (familial dysautonomia)*, a disease associated with degenerative changes in the central nervous system and the peripheral autonomic system. Symptoms include decreased lacrimation, transient skin blotching, attacks of hypertension, episodes of hyperpyrexia and vomiting, impairment of taste discrimination, relative insensitivity to pain, and emotional instability.

2. *Adiposogenital syndrome*, characterized by obesity and retarded development of secondary sexual characteristics, sometimes is associated with lesions in the hypothalamus.

3. Precocious puberty may result from hypothalamic tumors.

4. The common cold. Temperature elevation is apparently the consequence of some influence on hypothalamic functioning.

5. Tumors of the pituitary may have a destructive effect on the pituitary gland and hypothalamus by direct extension, e.g. in chromophobe adenoma and craniopharyngioma which generally are nonsecretory tumors. If the tumor contains functioning glandular tissue e.g., acidophilic or basophilic adenoma, there may be the opposite effect of hypersecretion of pituitary hormones.

6. *Diabetes insipidus.* Vasopressin (antidiuretic hormone) enhances the reuptake of water in the kidney. Interference with its production, as by an invading tumor, leads to diabetes insipidus, characterized by excessive production of urine and excessive thirst (up to 20 liters imbibed daily).

7. *Horner's syndrome.* Interruption of the cervical sympathetic nerves (or in some cases their central origins in the spinal cord and brain stem) leads to ptosis, miosis and decrease in sweating on the involved side of the face. Sometimes this is the result of a tumor of the apex of the lung (*Pancoast tumor*) that interrupts the fibers as they course from the superior cervical ganglion (the most rostral ganglion in the sympathetic chain) to the carotid artery on their way to the orbit.

Surgical procedures are performed that interrupt the sympathetic innervation of the lower extremities, in order to increase circulation in cases of vascular insufficiency.

8. *Hirschsprung's megacolon*—Congenital absence of parasympathetic ganglion cells in the wall of the colon, resulting in poor colonic motility and a dilated colon.

9. *Shy-Drager syndrome* (Multiple System Atrophy; MSA) is a condition, usually in men over 60 years old, that resembles Parkinson's disease (slowness, stiffness, mask-like facies, pill-rolling tremor) along with autonomic symptoms (fainting when standing up, loss of bowel/bladder control, impotence, decreased sweating, nausea and digestive problems).

Questions

6-1   Which type of fiber—sympathetic or parasympathetic is damaged by a lesion to the cauda equina at vertebral level L5?

        Ans.   Parasympathetic. The last sympathetic nerve root exits through its foramen at vertebral level L2. Parasympathetic nerves S2-S4 originate in

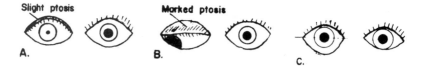

Slight ptosis   Marked ptosis

A.   B.   C.

Fig. 49   Answers to question 6-2 A. Miosis and slight ptosis. B. Dilated fixed pupil and marked ptosis. C. No pupillary changes. The interpalpebral fissure may be slightly wider on the right because of loss of muscle tone in the orbicularis oculi (CN7) muscle.

the spinal cord in the vicinity where the spinal cord ends—about vertebral level L1-L2. These parasympathetic fibers then travel down the cauda equina to exit at vertebrae S2-S4.

6-2   Draw the appearance of the eyes in A. A right Horner's syndrome. B. A right CN3 lesion. C A right CN7 lesion.
Ans.   Figure 49.

6-3   Why do patients with migraine headache sometimes develop a constricted pupil during the migraine attack?
Ans.   Edema along the carotid artery at the time of migraine may compromise the sympathetic nerves that travel with the carotid artery and its branches.

6-4   What is the *locus ceruleus?*
Ans.   It is a norepinephrine-containing brain stem nucleus that lies near the mesencephalic nucleus of CN5. It projects to widespread areas of the brain and may have a general effect on modulating brain function.

6-5   What is the *reticular formation?*
Ans.   It is simply, any area of grey matter that is unlabeled in diagrams. As more becomes known about it, more labels will appear and its size will diminish. It has important motor and sensory functions, including those relating to the autonomic nervous system (e.g., centers in the medulla controlling heart rate and blood pressure). Multisynaptic pathways through the reticular formation, from hypothalamus to spinal cord, convey sympathetic information. Thus, it is possible to acquire a Horner's syndrome from a brain stem lesion.

Endogenous chemicals with opiate-like activity (*endorphins*) and their receptors have been found in various areas of the reticular formation (e.g., the grey matter surrounding the aqueduct, and cells along the midline of the brain stem). The implication of the system in the relief of pain is currently a topic of great interest.

CHAPTER 7.   CEREBELLUM, BASAL GANGLIA, AND THALAMUS

Clinically, it is not very important to know the complex internal connections of the thalamus, cerebellum and basal ganglia, and these will not be emphasized.

The Thalamus

The thalamus is a sensory relay and integrative center connecting with many areas of the brain including the cerebral cortex, basal ganglia, hypothalamus and brain stem. It is capable of perceiving pain but not of accurate localization. For

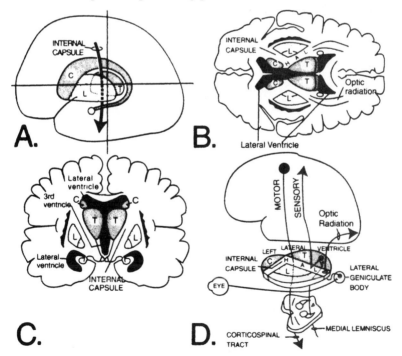

Fig. 50   The internal capsule and its relationship to the caudate nucleus (C), thalamus (T), and lentiform nucleus (L). A. Lateral view. B. Horizontal section at the level indicated in A. C. Cross section at the level indicated in A. D. The course of the major motor and sensory pathways. H, head; A, arm; L, leg.

instance, patients with tumors of the thalamus may experience the "thalamic pain syndrome"—a vague sense of pain without the ability to accurately localize it. Sensory fibers, ascending through the brain stem, synapse in the thalamus and are then relayed to the cerebral cortex via the internal capsule. Motor fibers descending from the cortex pass to the brain stem via the internal capsule without synapsing in the thalamus. Figure 50 illustrates the anatomy of this region.

Important thalamic nuclei include: the ventral posterolateral (VPL) nucleus, a synaptic region for ascending spinal sensory pathways (Fig. 15); the ventral posteromedial (VPM) nucleus, the synaptic area for the trigeminal lemniscus (Fig. 27); the anterior thalamic nucleus, which is a part of the Papez circuit (Fig. 47); the ventral lateral (VL) nucleus, which receives input from the cerebellum (Fig. 51).

The *lentiform nucleus* (the *putamen* plus *globus pallidus*) lies lateral to the internal capsule, and the caudate nucleus and thalamus lie medial. Note in Figure 50D the distribution of head, arm and leg fibers in the internal capsule.

The Cerebellum and Basal Ganglia

Rather than list the multitude of complex cerebellar and basal ganglia connections, it is clinically more important to understand the types of clinical syndromes that may occur in these two systems. In general, cerebellar dysfunction is characterized by awkwardness of intentional movements. Basal ganglia disorders are more characterized by meaningless unintentional movements occurring unexpectedly.

Cerebellar Disorders

1. Ataxia—awkwardness of posture and gait; tendency to fall to the same side as the cerebellar lesion; poor coordination of movement; overshooting the goal in reaching toward an object (dysmetria); inability to perform rapid alternating movements (dysdiadochokinesia), such as finger tapping; scanning speech due to awkward use of speech muscles, resulting in irregularly spaced sounds.

2. Decreased tendon reflexes on the affected side.

3. Asthenia—muscles tire more easily than normal.

4. Tremor—usually an intention tremor (evident during purposeful movements).

5. Nystagmus.

Basal Ganglia Disorders

1. *Parkinsonism*—rigidity; slowness; resting tremor; mask-like facies; shuffling gait, associated with degeneration in the basal ganglia and substantia nigra of the midbrain (Figs. 72, 73).

2. *Chorea*—sudden jerky and purposeless movements (e.g. Sydenham's chorea found in rheumatic fever; Huntington's chorea, an inherited disorder).

3. *Athetosis*—slow writhing, snake-like movements, especially of the fingers and wrists.

4. *Hemiballismus*—a sudden wild flail-like movement of one arm.

Questions

7-1   In the *Romberg* test the patient is asked to close his eyes while standing. If he sways back and forth with his eyes closed, but does not sway with them open, then the Romberg test is called "positive." Normally there should be no swaying even with the eyes closed. In what clinical conditions would you expect a positive Romberg?
Ans.   In proprioceptive or vestibular defects, for the following reason. To keep one's balance requires at least two out of the three senses that help maintain balance—vision, vestibular sense and proprioception. These three modalities feed into the cerebellum. Thus, if either proprioception or vestibular sense are defective, the patient will sway if he also closes his eyes. One can distinguish a proprioceptive from a vestibular deficit by neurological testing. In the proprioceptive defect the patient experiences difficulty in determining if his toes are being flexed or extended by the examiner. In vestibular defects he may experience vertigo (a sense of spinning of the patient or his environment), nystagmus or abnormal cold caloric testing.

7-2   Will a patient with a pure cerebellar defect have a positive Romberg?
Ans.   No—he will sway with or without his eyes closed and the definition of "Romberg" is quite strict (see question 7-1).

7-3   A *red nucleus* lesion results in tremor of the right arm and leg. Which red nucleus is involved—right or left?
Ans.   Left. The right cerebellum controls the right side of the body. Part of this mechanism involves fibers which enter the superior cerebellar peduncle and cross over to the left red nucleus. Fibers eventually cross back again (Fig. 51) to the right side. Hence a lesion to the left red nucleus results in a right-sided deficit.

7-4   A patient has right-sided paresis of the extremities and lower face, sensory loss (pain, touch, conscious proprioception) on the right side of the body, and right homonymous hemianopia. Where is the smallest lesion that would account for this?
Ans.   Posterior limb of the left internal capsule, including the optic radiation (see Fig. 50D). This is a common lesion, often from a stroke in this area (e.g., involving choroidal or striate arteries—Fig. 7).

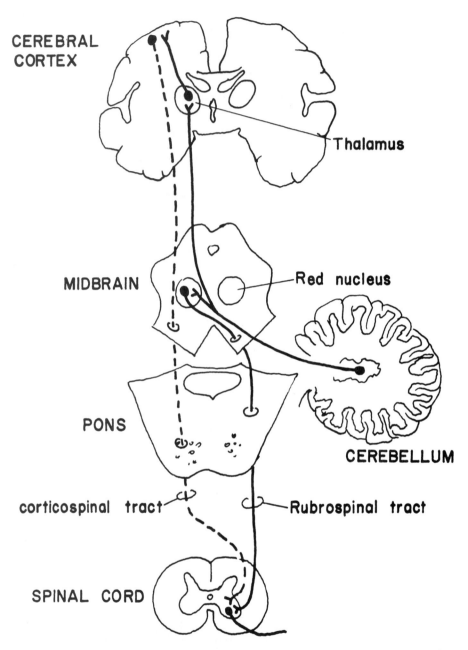

CEREBRAL
CORTEX

Thalamus

MIDBRAIN

Red nucleus

PONS

CEREBELLUM

corticospinal tract

Rubrospinal tract

SPINAL CORD

Fig. 51   The relationship between the cerebellum and the red nucleus. (Modified from Clark, R.G., *Manter & Gatz's Essentials of Clinical Neuroanatomy & Neurophysiology,* F.A. Davis Company, Philadelphia, 1975.)

CHAPTER 8. CEREBRAL CORTEX

Lesions to the nervous system may lead to simple or complex levels of dysfunction depending upon the area involved. For instance, if one asks a patient to put on a polo shirt and his left brachial plexus is severed, he will only use his right arm, as his left arm is paralyzed. The cerebellum and basal ganglia represent a step further in levels of functioning. With a cerebellar lesion, the patient may perform the act awkwardly, e.g. overshooting the mark, or with tremor. With basal ganglia lesions there may be unexpected, unplanned totally irrelevant movements, beyond the realm of awkwardness, e.g., a sudden flail-like movement, etc. In the cerebral cortex, unless the lesion is in the primary motor area wherein *paralysis* may result, a higher level of dysfunction may be found, e.g. trying to get his head into the sleeve or trying other inadequate orientations.

In speech, a lesion of CN10 may result in hoarseness (laryngeal dysfunction). When cranial nerves 10, 12, or 7 are involved, there may be difficulty with the "KLM" sounds: i.e., the sounds "Kuh, Kuh, Kuh" test the soft palate (CN10), "LA, LA, LA," the tongue (CN12) and "MI, MI, MI" the lips (CN7). With lesions in the speech areas of the cerebral cortex, the deficit may involve higher levels of speech organization—the deletion of words or inclusion of excessive or inappropriate words. In psychiatric disturbances the level of dysfunction is even higher with abnormalities in the entire pattern of thought organization.

It is similar for the sensory pathways. Simple *anesthesia* may result from lesions between the primary sensory cortex and the body periphery. In other brain areas, the patient may have difficulty in comprehending the incoming information. For instance, a lesion to the optic tract results in homonymous hemianopia. In lesions to cortical areas 18 and 19 the patient sees but may not recognize what he is looking at.

Complex cerebral receptive disabilities are called *agnosias*. Complex cerebral motor disabilities are *apraxias*. Often the two are difficult to distinguish.

When language function is involved the disabilities are termed *aphasias*. Aphasia may be receptive (i.e., reading, listening) or motor (i.e., writing, talking). For aphasia to occur, the lesion must be in the dominant hemisphere, which is the left hemisphere in right handed people and in many (but not all) left handed people.

The shaded area in Figure 52 indicates the region of the cerebral cortex involved with aphasia. Subdivisions of this area are logical. Lesions in the anterior part of this area, near the motor cortex, tend to result in expressive aphasia. Lesions more posterior, near the auditory and visual cortex, result in receptive aphasia. Lesions nearer the visual cortex result in inability to read (alexia). Lesions near the auditory cortex result in inability to understand speech.

Actually, difficulty in talking may result from both anterior or posterior lesions. Following lesions in the more anterior regions of the aphasic zone, speech disturbances tend to be nonfluent; the patient omits nouns and connector words like *but*, *or* and *and*. In the more posterior regions his words are plentiful or even excessive, but he crams into his speech inappropriate word substitutes, circumlocutions and neologisms—a word salad. Presumably, this is because the ability to speak also depends on the ability to understand what one is saying. Thus, if the aphasic area near the auditory cortex is involved, this will also result in a defect in speech.

Lesions to corresponding areas of the non-dominant hemisphere do not result in aphasia, but rather in visual or auditory inattention to the left environment or

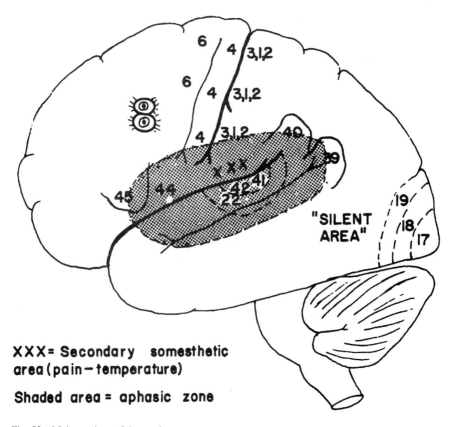

XXX= Secondary somesthetic area (pain—temperature)

Shaded area = aphasic zone

Fig. 52   Major regions of the cerebrum.

to general unawareness of the concept of "left." The patient may deny he has any neurological deficit despite a dense hemiplegia and left visual field defect.

Specific cerebral cortical regions and the effects of lesions are listed below. Area 4 (the primary motor area). Lesions result in initial flaccid paralysis followed in several months by partial recovery of function and a possible Babinski reflex; spasticity and increased deep tendon reflexes may occur if area 6 (the supplemental motor area) is included.

Lesions to area 8 (the frontal eye fields) result in difficulty in voluntarily moving the eyes to the opposite side.

Areas of the frontal cortex rostral to the motor areas are involved in complex behavioral activities. Lesions result in changes in judgment, abstract thinking, tactfulness and foresight. Symptoms may include irresponsibility in dealing with daily affairs, vulgar speech and clownish behavior.

Areas 44, 45 (Broca's speech area). The patient with a lesion in this area experiences motor aphasia, but only when the dominant hemisphere is involved. The patient knows what he wants to say but speech is slow, deleting many prepositions and nouns.

Areas 3, 1, 2 (primary somesthetic area). Lesions produce contralateral impairment of touch, pressure and proprioception. Pain sensation will be impaired if the lesion lies in the secondary somesthetic sensory area (Fig. 52) which receives pain information.

Areas 41, 42 (auditory area). Unilateral lesions have little effect on hearing owing to the bilateral representation of the auditory pathways. Significant auditory defects generally involve either CN8 or its entry point in the brain stem, for bilateral representation begins beyond this point.

Area 22 (Wernicke's area). Lesions in the dominant hemisphere result in auditory aphasia. The patient hears but does not understand. He speaks but makes mistakes unknowingly owing to his inability to understand his own words.

Area 40 (supramarginal gyrus). Lesions in the dominant hemisphere result in tactile and proprioceptive agnosia, and a variety of other problems, such as confusion in left-right discrimination, disturbances of body-image, and apraxia, by cutting off impulses to and from association tracts that interconnect this area with nearby regions.

Area 39 (angular gyrus). Lesions in the dominant hemisphere may result in alexia and agraphia (inability to read and write).

Areas 17, 18, 19. Total destruction causes blindness in the contralateral visual field. Lesions to areas 18 and 19 alone do not cause blindness but rather difficulty in recognizing and identifying objects (visual agnosia).

The silent area is believed to function in memory storage of visual and auditory information and is implicated in hallucinations and dreams. Epileptic attacks originating in this region may be associated with amnesia, auditory hallucinations, and the deja vu phenomenon.

The *basal nucleus of Meynert* lies in the base of the frontal lobe just lateral to the optic chiasm. Degeneration of this nucleus is associated with the dementia of Alzheimer's disease.

CHAPTER 9.   CLINICAL REVIEW

9-1   What general principles are useful in determining whether a lesion lies at
        the level of the cerebral cortex, internal capsule, cerebellum, basal ganglia,
        brain stem, spinal cord or peripheral nerve?
        Ans.   Cerebellar and basal ganglia lesions result in motor problems,
        specifically in aberrations in the quality of coordinated movements, as
        opposed to paralysis. Cerebellar dysfunction is characterized by awkward-
        ness of intentional movements. Basal ganglia disorders are more character-
        ized by meaningless, unintentional, unexpected movements.
            Lesions in the cerebral cortex and internal capsule both result in sensory
        and motor defects confined to the contralateral environment. It may be dif-
        ficult to differentiate a lesion in the cerebral cortex from one in the internal
        capsule. The presence of higher level dysfunction, particularly an agnosia
        or apraxia, is more consistent with a cerebral cortex lesion.
            Unilateral brain stem and spinal cord lesions result in ipsilateral as well
        as contralateral defects, owing to the crossing over of certain pathways
        and not others. In the spinal cord, a unilateral lesion results in ipsilateral
        paralysis and proprioceptive loss and contralateral pain-temperature loss
        below the level of the lesion. A unilateral brain stem lesion results in con-
        tralateral upper motor neuron paralysis and in contralateral proprioceptive
        and pain-temperature loss below the head, and in ipsilateral cranial nerve
        defects. The presence of cranial nerve involvement signifies that the lesion
        lies above the level of the foramen magnum. The presence of radicular
        pain along an extremity suggests that the lesion lies below the level of the
        foramen magnum, but it should be noted that such pain may be incidental
        to more peripheral problems. The presence of a cranial nerve defect on
        one side and defects of motor or sensory modalities in the contralateral
        extremities confirms that the lesion lies at the level of the brain stem and
        not in the cerebral cortex or internal capsule.
            Peripheral nerve injuries result in ipsilateral motor and sensory defects.
        Peripheral nerve lesions may be distinguished from internal capsule
        and cerebral cortical lesions by the presence of lower motor neuron signs and
        motor and sensory defects along a dermatome-like distribution (Figure 53).

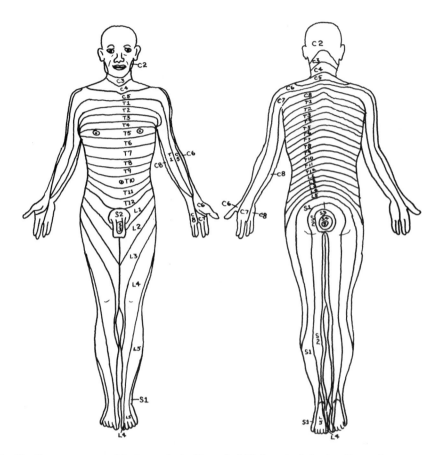

Fig. 53   Dermatome map of the human body. C, cervical; T, thoracic; L, lumbar; S, sacral.

In distinguishing a spinal nerve root lesion from a more peripheral nerve lesion, several points should be considered. Severing a single spinal nerve root commonly results in little if any motor or sensory defect owing to the overlap distribution of adjacent roots. It may be necessary to sever three or more roots to cause a significant motor or sensory defect. With certain exceptions the dermatome (skin) map of the various spinal nerve roots approximately overlies the muscular map distribution of the same spinal cord segments. Spinal nerve root lesions result in deficits that differ from those that follow lesions of peripheral nerve plexuses and more peripheral extensions of the nerve. This is a consequence of the fact that a peripheral nerve is a mixture of fibers arising from several nerve roots.

Figures 54 and 55 illustrate the characteristic motor and sensory deficits that arise from classical peripheral nerve lesions.

9-2   Localize the lesions in Figures 56-61 (shaded areas indicate regions of functional deficit).

MOTOR FUNCTIONS IMPAIRED WITH INJURY

NERVE	
RADIAL (C5-C8)	Elbow and wrist extension (patient has wrist drop); extension of fingers at metacarpo-phalangeal joints; triceps reflex.
MEDIAN (C6-T1)	Wrist, thumb, index, and middle finger flexion; thumb opposition, forearm pronation; ability of wrist to bend toward the radial (thumb) side; atrophy of thenar eminence (ball of thumb).
ULNAR (C8-T1)	Flexion of wrist, ring and small finger (claw hand); opposition little finger; ability of wrist to bend toward ulnar (small finger) side; adduction and abduction of fingers; atrophy of hypothenar eminence in palm (at base of ring and small fingers).
MUSCULOCUTANEOUS(C5-C6)	Elbow flexion (biceps); forearm supination; biceps reflex.
AXILLARY (C5-C6)	Ability to move upper arm outward, forward, or backward (deltoid atrophy).
LONG THORACIC (C5-C7)	Ability to elevate arm above horizontal (patient has winging of scapula).
FEMORAL (L2-L4)	Knee extension; hip flexion; knee jerk.
OBTURATOR (L2-L4)	Hip adduction (patient's leg swings outward when walking).
SCIATIC (L4-S3)	Knee flexion plus other functions along its branches—the tibial and common peroneal nerves.
Tibial (L4-S3)	Foot inversion; ankle plantar flexion; ankle jerk.
Common peroneal (L4-S2)	Foot eversion; ankle and toes dorsiflexion (patient has high slapping gait owing to foot drop). This nerve is very commonly injured.

Fig. 54

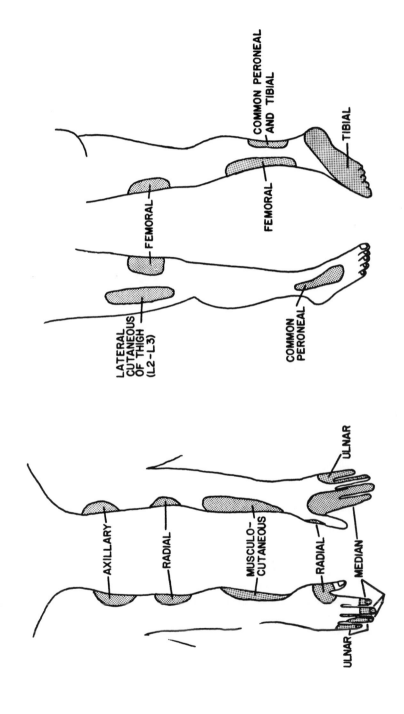

Fig. 55   Sensory defects following peripheral nerve injuries.

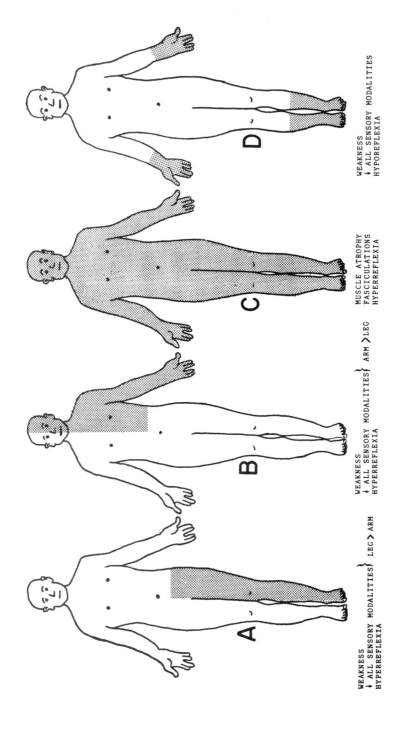

WEAKNESS
↓ ALL SENSORY MODALITIES} LEG > ARM
HYPERREFLEXIA

A

WEAKNESS
↓ ALL SENSORY MODALITIES} ARM > LEG
HYPERREFLEXIA

B

MUSCLE ATROPHY
FASCICULATIONS
HYPERREFLEXIA

C

WEAKNESS
↓ ALL SENSORY MODALITIES
HYPOREFLEXIA

D

Fig. 56

72

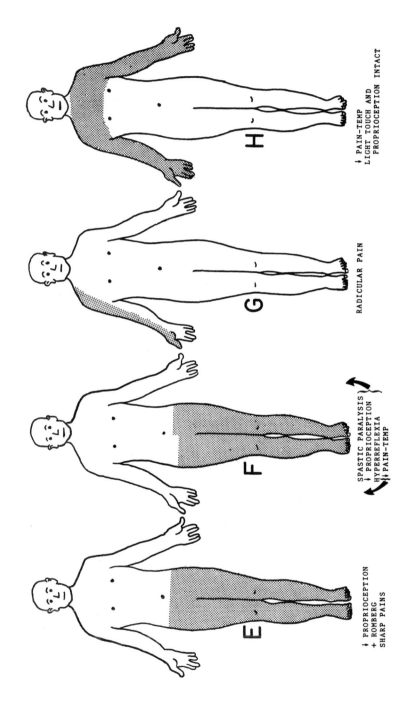

E
↓ PROPRIOCEPTION
+ ROMBERG
SHARP PAINS

F
SPASTIC PARALYSIS
↓ PROPRIOCEPTION
HYPERREFLEXIA
↓↓ PAIN-TEMP

G
RADICULAR PAIN

H
↓ PAIN-TEMP
LIGHT TOUCH AND
PROPRIOCEPTION INTACT

Fig. 57

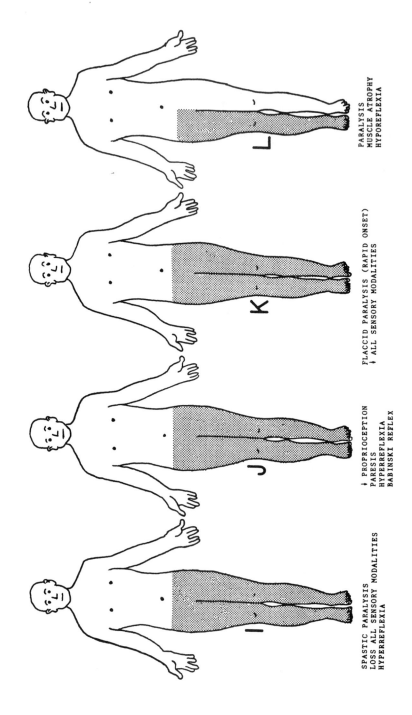

I

SPASTIC PARALYSIS
LOSS ALL SENSORY MODALITIES
HYPERREFLEXIA

J

↓ PROPRIOCEPTION
PARESIS
HYPERREFLEXIA
BABINSKI REFLEX

K

FLACCID PARALYSIS (RAPID ONSET)
↓ ALL SENSORY MODALITIES

L

PARALYSIS
MUSCLE ATROPHY
HYPOREFLEXIA

Fig. 58

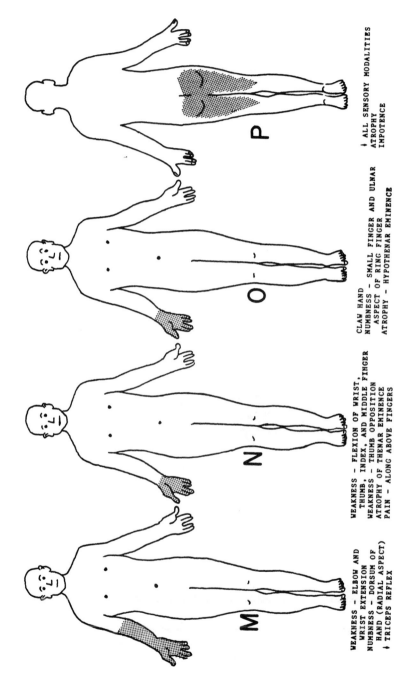

WEAKNESS – ELBOW AND
    WRIST EXTENSION
NUMBNESS – DORSUM OF
    HAND (RADIAL ASPECT)
↓ TRICEPS REFLEX

M

WEAKNESS – FLEXION OF WRIST,
    THUMB, INDEX, AND MIDDLE FINGER
WEAKNESS – THUMB OPPOSITION
ATROPHY OF THENAR EMINENCE
PAIN – ALONG ABOVE FINGERS

N

CLAW HAND
NUMBNESS – SMALL FINGER AND ULNAR
    ASPECT OF RING FINGER
ATROPHY – HYPOTHENAR EMINENCE

O

↓ ALL SENSORY MODALITIES
ATROPHY
IMPOTENCE

P

Fig. 59

75

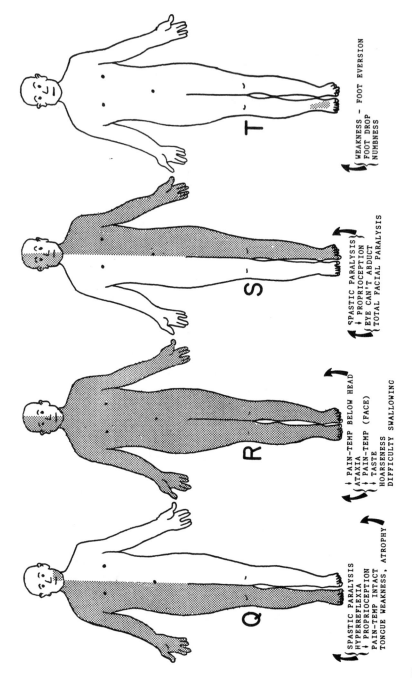

Q

SPASTIC PARALYSIS
↓ HYPERREFLEXIA
↓ PROPRIOCEPTION
PAIN-TEMP INTACT
TONGUE WEAKNESS, ATROPHY

R

↓ PAIN-TEMP BELOW HEAD
ATAXIA
↓ PAIN-TEMP (FACE)
↓ TASTE
HOARSENESS
DIFFICULTY SWALLOWING

S

SPASTIC PARALYSIS
↓ PROPRIOCEPTION
EYE CAN'T ABDUCT
TOTAL FACIAL PARALYSIS

T

WEAKNESS – FOOT EVERSION
FOOT DROP
NUMBNESS

Fig. 60

76

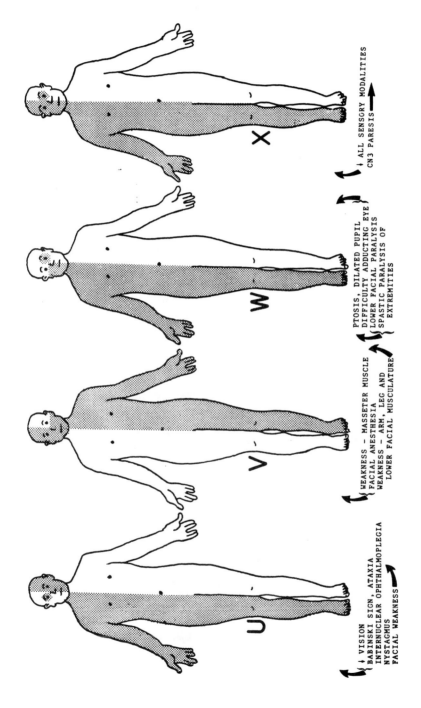

VISION
BABINSKI SIGN, ATAXIA
INTERNUCLEAR OPHTHALMOPLEGIA
NYSTAGMUS
FACIAL WEAKNESS

WEAKNESS - MASSETER MUSCLE
FACIAL ANESTHESIA
WEAKNESS - ARM, LEG AND
LOWER FACIAL MUSCULATURE

PTOSIS, DILATED PUPIL
DIFFICULTY ADDUCTING EYE
LOWER FACIAL PARALYSIS
SPASTIC PARALYSIS OF
EXTREMITIES

ALL SENSORY MODALITIES
CN3 PARESIS

U

V

W

X

Fig. 61

77

Ans. (Figs. 56-61).

A. Right anterior cerebral artery (see Figs. 7, 8).

B. Right middle cerebral artery (see Figs. 7, 8).

C. Amyotrophic lateral sclerosis (Lou Gehrig's disease) (see Fig. 18A). Distal extremities commonly affected first.

D. Peripheral neuropathy (see question 3-5B).

E. Tertiary syphilis (tabes dorsalis). Upper extremities less commonly affected (see Fig. 18B).

F. Hemisection of spinal cord (Brown-Sequard Syndrome) on left at T11 (see Fig. 15 and question 3-1).

G. Radicular pain, cervical roots C5-C6.

H. Syringomyelia C5-T2 (see Fig. 18F and question 3-3), Cervical levels are most commonly affected, often with atrophy of the small muscles of the hand.

I. Total transection of spinal cord, T11. The deficit could also result from a midline tumor at the level of the central sulcus, but the deficit in the latter would probably be less dense than that following a spinal cord transection.

J. Pernicious anemia. Commonly also presents with numbness and tingling of the distal portion of all extremities secondary to peripheral nerve involvement (see Fig. 18C).

K. Guillain-Barre syndrome. Commonly ascends to upper extremities and face after affecting lower extremities (see Fig. 18E).

L. Polio (see Fig. 18D).

M. Radial nerve injury.

N. Median nerve compromise. When injury is at level of wrist (e.g. carpal tunnel syndrome), wrist flexion is unaffected.

O. Ulnar nerve injury.

P. Cauda equina lesion, S2-S4.

Q. Lesion of left medulla (see question 5-3B).

R. Syndrome of the posterior inferior cerebellar artery (see question 5-3C). May also cause contralateral loss of facial pain-temp, if trigeminal lemniscus involved (see Fig. 27).

S. Lesion of the right caudal pons (see question 5-3F).

T. Common peroneal nerve injury.

U. Multiple sclerosis (no single lesion possible).

V. Lesion of right pons (see question 5-3H).

W. Lesion of left midbrain (see question 5-3I).

X. Lesion of left midbrain (see question 5-3J).

CHAPTER 10.   MINIATLAS

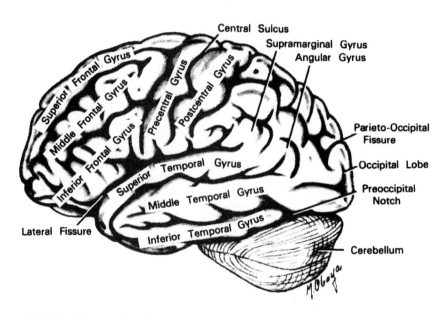

Fig. 62   Lateral view of the brain.

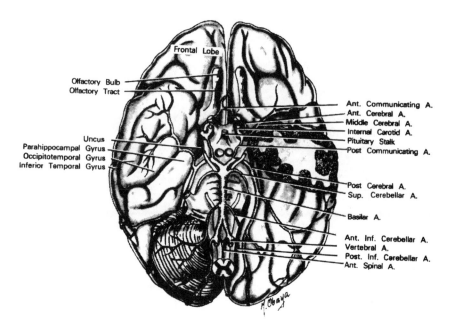

Fig. 63   Basal view of the brain.

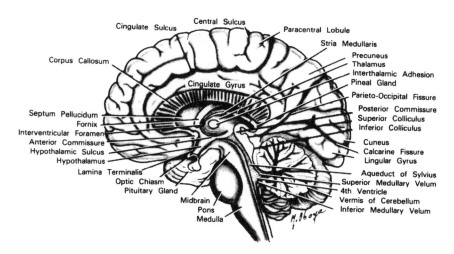

Fig. 64   Midsagittal view of the brain. The lateral wall of the 3rd ventricle, on each side of the midline, consists partly of thalamus and partly of hypothalamus.

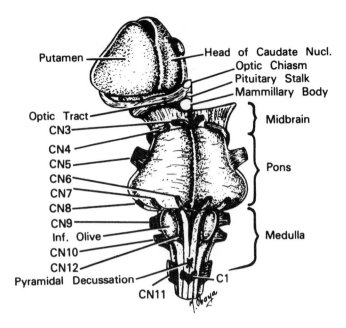

Fig. 65   Brain stem, basal (anterior) view.

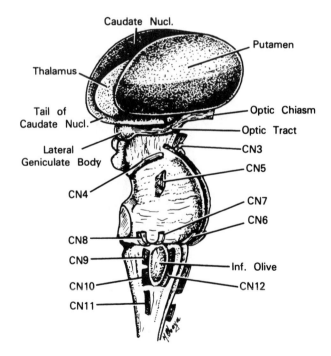

Fig. 66   Brain stem, lateral view. The internal capsule (not pictured) travels between the lentiform nucleus (= putamen + globus pallidus), on one side, and the thalamus and caudate nucleus on the other (see Fig. 50). The globus pallidus lies just medial to the putamen and is obscured by the putamen in Fig. 66.

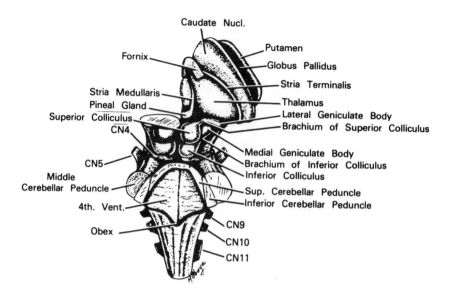

Fig. 67   Brain stem, dorsal (posterior view).

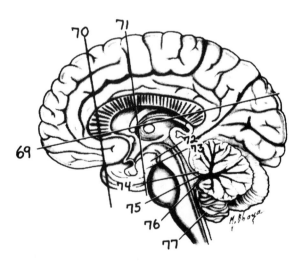

Fig. 68   Orientation of key sections through the brain in figures 69-77.

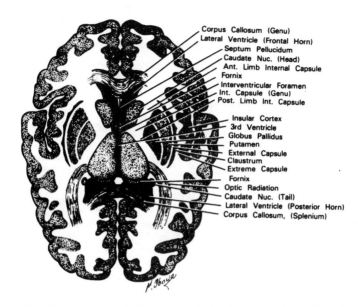

Corpus Callosum (Genu)
Lateral Ventricle (Frontal Horn)
Septum Pellucidum
Caudate Nuc. (Head)
Ant. Limb Internal Capsule
Fornix
Interventricular Foramen
Int. Capsule (Genu)
Post. Limb Int. Capsule

Insular Cortex
3rd Ventricle
Globus Pallidus
Putamen
External Capsule
Claustrum
Extreme Capsule
Fornix
Optic Radiation
Caudate Nuc. (Tail)
Lateral Ventricle (Posterior Horn)
Corpus Callosum, (Splenium)

Fig. 69   Horizontal section through the interventricular foramen and splenium of the corpus callosum.

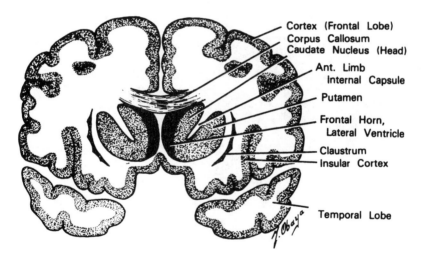

Cortex (Frontal Lobe)
Corpus Callosum
Caudate Nucleus (Head)
Ant. Limb
   Internal Capsule
Putamen
Frontal Horn,
   Lateral Ventricle
Claustrum
Insular Cortex
Temporal Lobe

Fig. 70   Coronal (cross) section placed anteriorly, through the area of fusion of the caudate nucleus and the lentiform nucleus' putamen (see fig. 50).

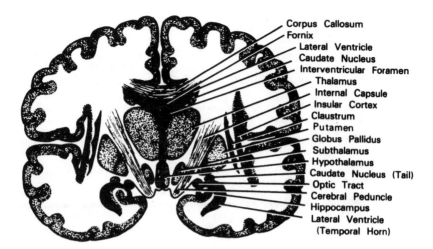

Corpus Callosum
Fornix
Lateral Ventricle
Caudate Nucleus
Interventricular Foramen
Thalamus
Internal Capsule
Insular Cortex
Claustrum
Putamen
Globus Pallidus
Subthalamus
Hypothalamus
Caudate Nucleus (Tail)
Optic Tract
Cerebral Peduncle
Hippocampus
Lateral Ventricle
(Temporal Horn)

Fig. 71 Coronal section through the interventricular foramen and thalamus, showing the extension of the internal capsule toward the cerebral peduncle of the midbrain.

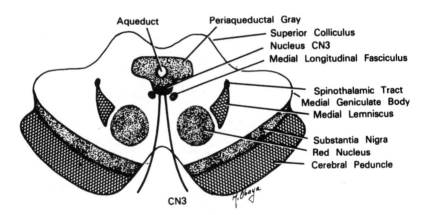

Aqueduct
Periaqueductal Gray
Superior Colliculus
Nucleus CN3
Medial Longitudinal Fasciculus
Spinothalamic Tract
Medial Geniculate Body
Medial Lemniscus
Substantia Nigra
Red Nucleus
Cerebral Peduncle
CN3

Fig. 72 Rostral midbrain.

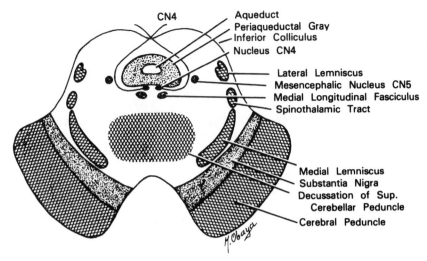

CN4
Aqueduct
Periaqueductal Gray
Inferior Colliculus
Nucleus CN4

Lateral Lemniscus
Mesencephalic Nucleus CN5
Medial Longitudinal Fasciculus
Spinothalamic Tract

Medial Lemniscus
Substantia Nigra
Decussation of Sup.
   Cerebellar Peduncle
Cerebral Peduncle

Fig. 73   Caudal midbrain.

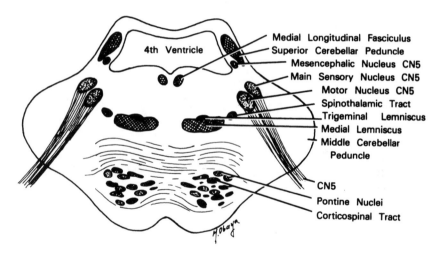

4th Ventricle

Medial Longitudinal Fasciculus
Superior Cerebellar Peduncle
Mesencephalic Nucleus CN5
Main Sensory Nucleus CN5
Motor Nucleus CN5
Spinothalamic Tract
Trigeminal Lemniscus
Medial Lemniscus
Middle Cerebellar
   Peduncle

CN5
Pontine Nuclei
Corticospinal Tract

Fig. 74   Rostral pons.

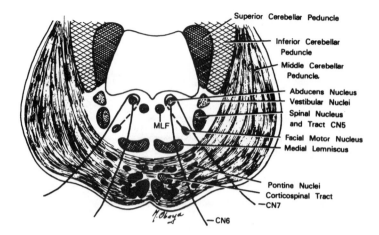

Fig. 75   Caudal pons. One can see all three cerebellar peduncles in this one section. MLF, medial longitudinal fasciculus.

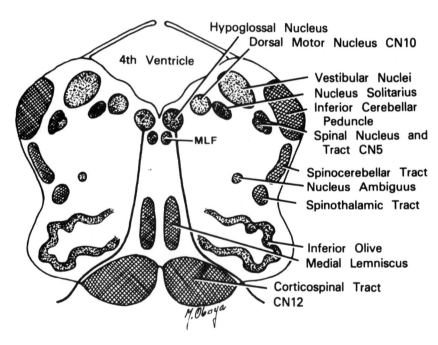

Fig. 76   Rostral medulla. MLF, medial longitudinal fasciculus.

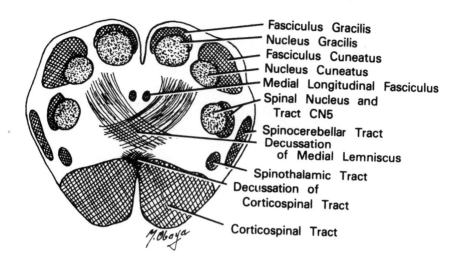

Fig. 77 Caudal medulla.

CHAPTER 11.   NEUROTRANSMITTERS

Apart from the anatomy of the nervous system, some mention should be made of neurotransmitters, which significantly affect the action of the nervous system.

Neurotransmitters are chemicals that are released from the ends of nerve cell axons and interact with postsynaptic cell receptors, either stimulating or inhibiting the firing of nerve cells. Decreases or increases in neurotransmitter function can significantly influence nervous system activity.

While some neurotransmitters are referred to as "inhibitory" or "excitatory," this classification is a bit misleading, since neurotransmitters are neither intrinsically inhibitory or excitatory. Their action depends on their receptors; some kinds of receptors can be stimulated, while others can be inhibited by the same neurotransmitter. Thus, a given neurotransmitter can be both inhibitory and excitatory, depending on the receptor. Moreover, even if a neurotransmitter excites a receptor and causes postsynaptic nerve cell firing, the firing cell may inhibit the next neuron in the line, and the neurotransmitter has a net inhibitory effect. Consider the substantia nigra and Parkinson's disease:

The *substantia nigra* of the midbrain produces the monoamine *dopamine*. The substantia nigra sends dopaminergic fibers to the striatum (= caudate nucleus + putamen), which then, via several intermediate synapses, influences the cerebral cortex by a "Direct" and an "Indirect" neural pathway (see figure below).

The "Direct" pathway stimulates, while the "Indirect" pathway inhibits motor regions of the cerebral cortex (Panel A in figure). Normally, there are two dopaminergic pathways from the substantia nigra to the striatum, one of which *stimulates the Direct pathway* while the other one *inhibits the Indirect pathway* (Panel B in figure). The net result is stimulation of the motor pathways of the cerebral cortex. In *Parkinson's disease*, though, there is degeneration of the substantia nigra and its dopamine pathways (Panel C in figure). This results in a net inhibition of the motor regions of the cerebral cortex, with the consequent bradykinesia (slow movement) and rigidity of Parkinson's disease. Treatment of Parkinson's disease aims to restore dopamine directly with L-dopa or stimulate dopamine receptors with agonists (drugs that mimic the neurotransmitter, e.g. *bromocriptine*), or decrease dopamine breakdown (e.g. *selegeline*). Surgical lesions of the inhibitory pathway may be an option.

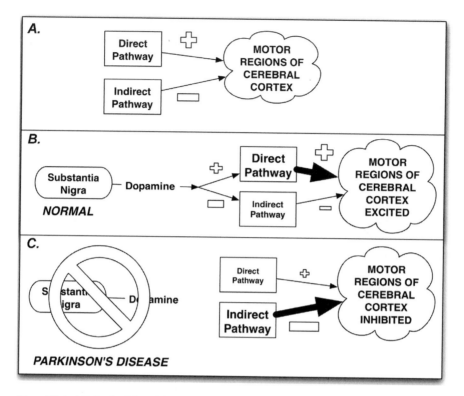

A.

Direct Pathway ➕ → MOTOR REGIONS OF CEREBRAL CORTEX

Indirect Pathway ➖

B.

Substantia Nigra — Dopamine → Direct Pathway ➕ → MOTOR REGIONS OF CEREBRAL CORTEX EXCITED

Indirect Pathway ➖ ➖

NORMAL

C.

Substantia Nigra — Dopamine

Direct Pathway ➕ → MOTOR REGIONS OF CEREBRAL CORTEX INHIBITED

Indirect Pathway ➖

PARKINSON'S DISEASE

(From Clinical Pathophysiology Made Ridiculously Simple, by A. Berkowitz, Medmaster, 2007)

Dopamine is low in Parkinson's disease and is believed to be high in schizophrenia. Antipsychotic drugs may therefore produce Parkinsonism as a side effect, due to their decrease of dopamine transmission.

Neurotransmitters are scattered throughout the nervous system, and, while some may be more confined to particular regions or nuclei, knowing the anatomy of these regions and nuclei is not as important as understanding the effects of a general decline or increase of the neurotransmitter, which may be more widespread than in a single isolated area. Thus, it is generally more important to know the effects of a general decline or increase in availability or function of a particular neurotransmitter than its particular location in the brain. Since neurotransmitters are widespread in the brain, affecting a neurotransmitter may have widespread effects, whether good or bad.

Too much neurotransmission can be caused by drugs that either increase the release of neurotransmitters, inhibit degradation of neurotransmitters, or act as agonists (mimics) at postsynaptic receptors. Too little neurotransmission can be caused by hereditary or environmental factors.

Acetylcholine and *norepinephrine* are the main neurotransmitters in the peripheral nervous system, which includes the cranial nerves, spinal nerves, and (discussed in Chapter 6) the autonomic nervous system. Acetylcholine (Ach), in addition to its autonomic function, stimulates receptors at the neuromuscular junction of skeletal muscles, enabling muscle contraction. *Myasthenia gravis* is an autoimmune disease where antibodies block skeletal muscle Ach receptors in skeletal muscles, resulting in muscle weakness. It is treated with *physostigmine*, an anticholinesterase, which prevents the breakdown of Ach, thereby keeping more Ach around. *Curare*, the "arrow poison," blocks the action of Ach at the neuromuscular junction, resulting in paralysis.

Acetylcholine is an important transmitter in the central nervous system as well, where it may have both excitatory and inhibitor receptors and is important in memory. Alzheimer's disease may be associated with a deficiency of acetylcholine in certain areas of the brain and is treated with acetylcholinesterase inhibitors.

The monoamines *serotonin, norepinephrine* and *dopamine*, when deficient, are believed important in the genesis of depression. Thus, drugs to help combat these disorders include:

- Selective serotonin reuptake inhibitors (SSRIs), e.g. *Prozac, Lexapro.*
- Norepinephrine reuptake inhibitors (NRIs), e.g. *Strattera.*
- Combined serotonin-norepinephrine reuptake inhibitors (SNRIs), such as the tricyclic antidepressant *Elavil* and newer drugs such as *Effexor* and *Cymbalta.*
- Combined dopamine-norepinephrine uptake inhibitors, e.g. *Wellbutrin.*
- Monoamine oxidase inhibitors (MAOIs), e.g. *Nardil, Parnate.* MAO is a family of enzymes that break down monoamines, such as serotonin, norepinephrine and dopamine, so MAO *inhibitors* can be useful in treating depression.

Dopamine and amphetamine excess produce a feeling of increased energy, less need for sleep, and an accelerated sense of time; they can also can lead to hallucinations and paranoid thinking. Mania (a hyperexcited mood and energy state) is believed related to overactivity of certain dopamine receptors. Cocaine increases dopaminergic activity. Marijuana reduces dopaminic activity and slows the feeling of subjective time.

Stimulant drugs that increase dopamine in the brain (e.g. *Ritalin*) or have amphetamine-like effects (e.g. *Adderall*) have been useful in treating *ADHD (Attention Deficit Hyperactivity Disorder)*, enabling patient to increase their focus.

Phenylethylamine (PEA) is a neurotransmitter in the brain that is associated with sexual attraction and excitement, peaking in orgasm and ovulation. It releases norepinephrine and dopamine. PEA is found in chocolate, but (unfortunately) is quickly metabolized when taken orally so that significant amounts do not reach the brain.

Glycine is an inhibitory transmitter. *Strychnine* poison is a glycine antagonist; it binds to the glycine receptor, resulting in spinal hyperexcitability, with severe muscle spasms, convulsions, and death.

- *GABA* ("Nature's Valium") is an inhiBitory amino acid neurotransmitter (in 90% of cases). *Benzodiazepines* and *barbiturates* enhance the actions of certain GABA receptors and are used as sedatives to treat anxiety, seizures, and the hyperactive muscle action of people with Huntington's disease.

- *GlutamaTe (glutamic acid)* is the main exci*T*atory neurotransmitter in the brain. Glutamate toxicity has been implicated in the development of *Huntington's chorea, Alzheimer's Disease,* and *amyotrophic lateral sclerosis (ALS).* Alertness due to *caffeine* ingestion may be related to its effect on increasing glutamate levels in the brain.
- *Opioids.* Opioid receptors are found in a number of areas in the nervous system. *Endorphins, enkephalins,* and *dynorphins* are natural body chemicals that interact with opiate receptors. They are powerful natural pain relievers. *Opium* and *morphine* mimic the natural opioids by binding to opioid receptors, thereby reducing pain. Opioid overdose can be treated with antagonists to opioid receptors, such as *naloxone (Narcan).*
- *Substance **P*** and *CGRP (Calcitonin Gene-Related Peptide)* are widespread neurotransmitters that among other things, are involved in the *production* of ***P*ain,** as in *migraine.* Medications that reduce levels of substance P and CGRP can be used as analgesics. *Neuropeptide Y* is a neurotransmitter that promotes increased food intake and is believed involved in the *reduction* of chronic pain.

Following injury, severed axons of the peripheral nervous system can regenerate, growing back to their targets at a rate of about 1-2 mm/day. The central nervous system, though, does not regenerate well, but motor and sensory function can improve through the brain's self-modification of alternative neural pathways and the use of a variety of mechanical and digital devices. Another approach is to pharmacologically enhance neurotransmitter activity.

GLOSSARY

acoustic neuroma — a tumor of the Schwann cell elements of the auditory nerve.

anesthesia — absence of sensation.

aneurysm — a focal ballooning out of a segment of blood vessel wall.

Babinski reflex — an abnormal reflex in which the great toe moves upward and the toes fan upon stroking the lateral plantar surface of the foot. It generally signifies an upper motor neuron lesion in an adult, but is normal in an infant.

basal ganglia — caudate nucleus, lentiform nucleus (= putamen plus globus pallidus), claustrum and amygdala.

bilateral — both sides of the body.

bitemporal hemianopia — loss of the temporal (lateral) visual field in each eye.

brachial artery — the major artery to the upper extremity, arising from the subclavian artery.

Brodmann's areas — a classification of areas of the cerebral cortex according to assigned numbers.

calcarine fissure — the horizontal fissure on the medial aspect of the occipital lobe. It separates the projections from the superior retinae (above) and the inferior retinae (below).

cauda equina — the bundle of nerve roots extending caudally from the caudal end of the spinal cord.

cerebral cortex — the narrow zone of grey matter that lies at the surface of the cerebrum. The grey matter lies external to the white matter in the cerebrum. The reverse is true for the spinal cord.

circumlocution — the use of many words to express what might be stated by few or one.

coma — depressed consciousness to the degree of unresponsiveness to noxious stimuli.

conductive hearing loss — hearing deficit resulting from a mechanical defect in the transmission of sound information between the external ear and the neuronal sensory apparatus (as opposed to neuronal hearing loss).

conjugate gaze — symmetrical movements of the eyes in a given direction.

contralateral — the opposite side of the body.

converge — to move the eyes toward one another.

corpus callosum — the major connecting pathway between left and right hemispheres.

corticobulbar tract — the motor pathway between the motor area of the cerebral cortex and the brain stem nuclei.

deja vu phenomenon — the feeling of having previously experienced a current event.

diencephalon — thalamus, hypothalamus, epithalamus, plus subthalamus.

dermatome — the area of skin supplied by one nerve root (Fig. 53).

dilation — expansion.

fasciculations — coarse muscle twitching seen with peripheral motor nerve injury.

fibrillations — fine, rarely visible, twitching of single muscle fibers, seen with peripheral motor nerve injury.

flaccid paralysis — paralysis in which the affected muscles are limp, with little resistance to passive movement.

hemiplegia — paralysis of the extremities on one side of the body.

homonymous hemianopia — loss of half of the visual field in each eye for a given side of the environment.

homunculus — the upside down representation of the human body on the cerebral cortex.

Horner's syndrome — pupillary constriction, slight ptosis, and decreased sweating resulting from interruption of the sympathetic pathways to the eye.

hyperpyrexia — elevation of body temperature.

hyperreflexia — overactive reflexes.

hyporeflexia — underactive reflexes.

infarction — a region of tissue death, resulting from obstruction of the local circulation.

intercostal arteries — the arteries that run between the ribs. These give contributory branches to the spinal cord.

intermediolateral columns — the autonomic zone of spinal cord grey matter, which lies between the anterior and posterior grey horns.

interpalpebral fissure — the space between the upper and lower eyelids.

ipsilateral — the same side of the body.

lamina terminalis — an emotion-related pathway connecting amygdala with hypothalamus.

lesion — injury.

lethargy — drowsiness.

medially — toward the midline.

meninges — the outer lining of the central nervous system (pia, arachnoid, and dura).

meningitis — inflammation of the meninges.

mesencephalic nucleus of 5 — the rostral most part of the sensory nucleus of CN5.

miosis — constriction of the pupil.

nasolabial fold — the skin crease extending from the nose to a point lateral to the corner of the mouth.

neologism — the use of a new word or an old word in a new sense.

neuron — nerve cell.

neuronal hearing loss — hearing deficit resulting from damage to CN8.

obex — the caudalmost landmark of the fourth ventricle.

obtundation, stupor — degrees of depressed consciousness between lethargy and coma.

olfaction — the sense of smell.

olive — the inferior olivary nucelus in the medulla.

ossicles — the small ear bones that transmit sound.

paresis — partial weakness, short of paralysis (which is total).

periaqueductal gray — the gray matter around the aqueduct, containing high concentrations of endorphins (opioid peptides).

posterior columns — the fasciculus gracilis plus fasciculus cuneatus.

pretectum — the area just deep to the superficial regions of the superior colliculus.

proprioception — the ability to sense the position of the limbs and their movements, with the eyes closed.

ptosis — drooping of the eyelid.

pyramidal decussation — crossover site of the corticospinal tracts.

radicular pain — pain along the distribution of a nerve root or primary nerve trunk.

reflex — an involuntary motion resulting from a stimulus (e.g. biceps jerk in response to percussion of the biceps tendon).

retina — the light-sensitive neural membrane within the eye.

retrograde — in reverse direction.

sagittal — in a plane parallel to that which divides an animal into right and left halves.

septum pellucidum — the thin midline membrane, between right and left lateral ventricles.

spastic paralysis — paralysis with a coinciding steady and prolonged involuntary contraction of the muscles affected.

stereognosis — the ability to recognize objects by touching and handling them with the eyes closed.

stria medullaris — a limbic system pathway on the dorsomedial surface of the thalamus, connecting septal to habenular nuclei.

stroke — a prolonged or permanent loss of function in a brain area, resulting from interruption of the blood supply.

subthalamus — nuclear region just lateral to the hypothalamus; lesions may result in hemiballismus (violent, flailing movements of a limb).

tympanic membrane — the vibratory membrane separating the external and middle ear.

uncus — a portion of the temporal lobe concerned with the sense of smell. Epileptic seizures in this area are characterized by unpleasant smells. The uncus lies close to CN3 and may press upon and injure it following a subdural hemorrhage that forces the uncus to herniate against the brain stem.

unilateral — on one side.

vestibular apparatus — the balance-sensing mechanism in the inner ear.

visual field — the portion of the environment that the eye(s) sees on fixed forward gaze.

INDEX